E Bonavia

Studies in the Evolution of Animals

E Bonavia
Studies in the Evolution of Animals
ISBN/EAN: 9783337229993

Printed in Europe, USA, Canada, Australia, Japan

Cover: Foto ©berggeist007 / pixelio.de

More available books at **www.hansebooks.com**

STUDIES IN THE
EVOLUTION OF ANIMALS

BY

E. BONAVIA, M.D.

BRIGADE SURGEON I.M.D.

AUTHOR OF 'THE CULTIVATED ORANGES AND LEMONS
OF INDIA AND CEYLON,' 'PHILOSOPHICAL NOTES ON
BOTANICAL SUBJECTS,' 'THE FLORA OF THE ASSYRIAN
MONUMENTS AND ITS OUTCOMES'

WITH ONE HUNDRED AND TWENTY-EIGHT ILLUSTRATIONS

Westminster
ARCHIBALD CONSTABLE AND COMPANY
PUBLISHERS TO THE INDIA OFFICE
14 PARLIAMENT STREET, S.W.
MDCCCXCV

[*All Rights reserved*]

CONTENTS

	PAGE
PREFACE,	vii
INTRODUCTION,	xv
PART I.—Spotted and Striped Mammals (Horses excepted),	1
,, II.—Dappled and Striped Horses, and some other Mammals,	57
,, III.—Meaning of the Jaguar and Leopard Rosettes, and of the markings of other Mammals,	99
,, IV.—Further evidence in support of the theory that existing Mammals descended from carapaced ancestors,	135
,, V.—Researches and discussions to connect, more surely, Armour-plating with Skin-picturing,	153
,, VI.—Probable meaning of some interesting features in Horses and other animals,	165
,, VII.—Is Natural Selection the sole factor in the Coloration of Mammals?	189
,, VIII.—Probable cause of the loss of the Calcareous Armour in Mammals,	199
,, IX.—Relationship between the Armadillo, the Rhinoceros, the Horse, the Giraffe, and the Zebu,	213

	PAGE
PART X.—Explanation of the Callosities on the Legs of Equine Animals and others,	229
,, XI.—The One Big Digit of the Horse,	241
,, XII.—Monstrosities as probable factors in the creation of species, .	273
GENERAL CONCLUSIONS, .	325
APPENDICES,	331

PREFACE

Of what value would the objective facts *alone* have been, even if they were collected by millions, *without* a colligating theory which puts *soul* into the scattered facts? They would be as soulless as heaps of bricks and mortar are before they are built up into a Cathedral. Theory founded on facts, as it should be, is of the greatest importance in forming a right conception of Nature. Theories cannot flash out in all perfection. They require to be mended, and time is needed for that.

PREFACE

THE genesis of these studies was the following :—

Having completed the *Flora of the Assyrian Monuments and its Outcomes*, I was looking about for something to take up next as a subject of study. In the furriers' windows I was attracted by the Leopard and Tiger skins, which by degrees became objects of interesting study and speculation.

Thinking over the rosettes of the Leopards, and more especially those of the Jaguar, and seeing spotted Horses constantly in the streets of London, some new ideas flashed across my mind regarding the origin of all this spotting and rosetting in mammals.

Keeping 'my eyes open,' so as to get some insight into the invisible, and making many researches for some tangible facts which would serve as a basis for my speculations, the subjects of the curious callosities on the legs of the Horse, its solitary big digit, its possible close relationship to the pair of digits in the ruminants, and various monstrosities, came under review, as well as several other collateral points, and so, each group, as it became interesting, was worked up into a separate study, followed up by that on the *meaning* of the rosettes on the Jaguar and allied animals, and of the dapples which are all but universal in Horses

—whether as a completely dappled surface, or as vestiges of an extensively spotted skin. Spotting and striping in mammals, or vestiges of such markings, are to be met with so extensively among these animals, that I came to the conclusion they must have a deeper meaning than may have hitherto been attributed to them by evolutionists.

The reader should understand that a book like this is not written as a poem might be written, by sitting down in some suggestive surrounding, gazing into space, and letting the thoughts come rambling after each other, as if by inspiration.

Not a few perhaps may think that the author just sat down and wrote it off! Few would consider what interminable searching for authorities was needed; what hunting for facts and evidence to build upon; what hunting for suitable skins and animals to be photographed, and for photographs of animals to be used as illustrations, were required. Nowadays, one might as well speak to the wind as produce a book of this sort without numerous illustrations. The value of an illustration is that it appeals to the mind at once, while a string of words used in a description, without an illustration, would only, in most cases, fatigue the reader's mind, and leave little or no impression. Everything must be made as easy as possible for the student and general reader, otherwise he or she will turn to something else. There is too much to distract the attention from making a serious effort to comprehend even a small portion of the work of creation. Moreover, this is the age of magazines, which mean a conglomerate, and most people prefer that.

The arrangement of notes taken at all sorts of odd times and places; the digesting and comparing of points, writing out

PREFACE xi

notions, tearing them up, and re-writing them, and a hundred other troubles, would all have been the most wearisome work had it not been backed by the stimulus and enthusiasm roused by the conviction that there was something interesting to be told. If the reader should feel a hundredth part of the interest, in reading these pages, that I felt in writing them, I am sure he ought to be a happy individual.

In these pages there may be some things which scientists may have either overlooked as unimportant, or which they may not have cared to tackle, appearing to them as insoluble.

I have also attempted to develop particular points in the sub-theories of the more general theory or doctrine of evolution. As Professor Huxley is stated to have declared at Oxford not long ago,[1] even if Darwinism were swept away it would leave evolution untouched. The doctrine of 'creation by the method of evolution' has replaced all other doctrines, and it cannot be upset by mortals, for it is based on the every-day experience that a mother procreates children, and these other children, and these others, and so forth. No one yet has even attempted to upset *that* fact of nature.

The study of the coloration of mammals is an intricate one; and in the various parts of this discussion repetitions could hardly be avoided; but although wearisome to the expert, the general reader, who may not be very conversant with the facts of evolution and natural selection—should he care to study this part of modern philosophy—may derive benefit from these

[1] After the Inaugural Address of Lord Salisbury to the British Association, 8th August 1894.

unavoidable though wearisome repetitions. Moreover, I am not aware that there is any particular sin in a little repetition. The reader after all is not so sacred a thing as not to be subjected on any account to a little tedium. Some little allowance then, I suppose, *may* be made for the 'personal equation' of the writer!

I do not know a better method for fixing notions on the reader's convolutions—more especially if he or she is not easily convinced—than by 'hammering' on the same subject in different ways, in order to make an impression. Few people who may be tempted to open a book do so with the spirit of the student who endeavours to master the meaning and points therein contained. The majority of persons who take up a book want to be amused, distracted, or somehow entertained, and few are the books on evolutionary studies which can satisfy either of these cravings.

Inferences cannot be safely drawn from any particular specimen; the larger the number of specimens on which an inference is based, the more soundness will it be likely to possess.

Where too much detail might seem tedious to the expert, I would note that it is intended for the general reader who may be tempted to dip into these subjects, and who may not have given much attention to such matters.

If found tedious, whole pages may be skipped by the expert.

I believe that this is the first time that any attempt has been made to study the markings of mammals in *detail*, with the view of reaching what seems to be the real, or at all events the proximate, cause of their existence. Mr. Tylor and Mr. Poulton,

PREFACE

Dr. Wallace and Mr. Darwin have studied the coloration of animals; but, as far as I am aware, no attempt has been hitherto made to account for certain markings which occur, as one might say, in a sort of *plan*, and in so many different animals. In these pages I have made an attempt to account, not only for their derivation, but also for their genesis, as far as this can be known.

I need hardly mention that the figures of Horses were not selected for their beauty of outline, but for their dappling. Most of the outline drawings are reduced in size.

My thanks are due to Professor J. M'Fadyean and Mr. P. D. Coghill of the Royal Veterinary College, for helping me with photographs of Horses; and to Mr. James Poynter of the Horse department of the Great Northern Railway, and Mr. William D. Duff, manager of the London Road Car Company, for their kind permission to photograph some of the horses in the stables of their companies. I have to thank also Messrs. Jeffs and Harris, and Messrs. Back and Co., of Regent Street, for allowing me to have some of the skins in their stores photographed. Finally, my thanks are due to Miss Butcher for numerous outline drawings of mammals, etc., in the Natural History Museum. Some apology is, I think, due to her for asking her to make drawings of those archæological exhibitions of Taxidermy. As she remarked, it was difficult to know where their legs ought to be! The *recent* specimens in that museum are, however, *splendidly got up*.

E. BONAVIA.

INTRODUCTION

'Evolution is the most striking feature of modern scientific thought, hence all that terms itself evolution must be scientific—such seems to be the logic of the average reviewer, and, we regret to say it, of some men of science who ought to know better. The fact is that the word "evolution" has been so terribly abused, first by the biologists, then by pseudo-scientists, and lastly by the public, that it has become a cant term to cover any muddle-headed reasoning, which would utterly fail to justify itself had it condescended to apply the rule of three. A variety of ill-described and ill-appreciated factors of change have all been classed together and entitled the theory of evolution.'

Socialism and Natural Selection, by KARL PEARSON,
Fortnightly Review, July 1894, p. 1.

ERRATA.

p. 49, Note I., for Vol. II., read Plate II.
p. 123, for *in* Fig. 62 (c), read Fig. 62 (c).
p. 132, for *ages* in the 4th para., read *stages*.
p. 192, for Fig. 73, read 76A.
p. 193, for Fig. 73, read 76A.
p. 287, for Gelasimus *arenatus*, read *arcuatus*.

INTRODUCTION

'IT is not enough that a scientific truth should be the possession of a privileged few; those who value the truth should try to spread it, and make it common intellectual property, and this can only be done when they realise that simplicity of language, and correct style, and a good arrangement, are essential to its propagation.'
'The British Association,' *Nature*, 16th August 1894.

ANY one who may think of devoting his attention to the study of the philosophy of life, based, not only on the materials he has himself observed and discovered, but also on those worked up by others, is met, at the outset, with a huge mountain of *words*. 'Words will govern us, if we do not govern them,' said Professor Max Müller. Any one who tries to get at the bottom of facts, and at the bottom of the inferences resulting from those facts, has to grope his way through this maze of often utterly useless, if not mischievous, terminology. The essential truth may be obscured by the novel and difficult-to-be-remembered wording; so that we often 'cannot see the wood for the trees.'

By way of introduction to the introduction, let us take a glance at what even leading scientists think of it all. If scientific men complain of the nuisance, what would you say the man in the street would think of it?

M. L. Guinard in his *Précis de Tératologie*, p. xvi., gives expression to the feeling of distress caused by hasty and reckless additions to modern scientific nomenclature. He says: 'La multiplication des termes aurait fatalement la même conséquence que la multiplication des langues ; celui qui songera à jeter les bases d'un édifice taxonomique par trop nouveau aboutira à l'édification d'une tour de Babel.

'Si j'insiste sur cette particularité, c'est que moi-même j'ai été souvent embarrassé par la diversité de noms et des classifications, et parce que actuellement, on voit la tendence dont je parle persister encore dans quelques travaux fort remarquables d'ailleurs.'

Then Mr. Stebbing, in his interesting work on the *Crustacea*, p. 255, referring to Mr. Spence Bates' 'Report on the Challenger *Macrura*,' says, 'But simplicity seems to be the very last thing considered in Spence Bates' terminology, and though such words as phymacerite, psalistoma, and stylamblys, may help to curtail the length of descriptions, they are only too likely also to curtail the number of those that read them.'

And certainly this is one of the mischiefs wrought by unnecessary coining of new terms to express ideas which might in many cases be conveyed in ordinary wording.

Very recently another note of warning has been sounded in *Natural Science* of October 1893, under the heading of 'Scientific Linguistics': 'When a layman asks a naturalist why he invents and employs such a multitude of incomprehensible technical terms, the common reply is that exact ideas necessitate a precise and universally (!) understood nomenclature. We wonder how this explanation would apply to the terms of " Auxology," or " Bioplastology," just discussed by Professor A. Hyatt in the *Zoologischer Anzeiger* (concluded August 28, 1893). We should like to know how much scientific precision there is in the determination of the nepionic, metanepionic, gerontic, paragerontic, etc., stages of any organism, and what grain of solid fact, as compared with mere speculation, in the so-called definition of the phylonepionic, phyloneanic, phylogerontic, etc., phases of development in any group of animals.

We may be enslaved by some prejudice, and our patience may have been ruffled in the attempt to decipher some recent writings of American authors on fossil shells ; but we cannot help uttering a protest against the clothing of a tissue of hypothetical fabrications in a garb of a precisely-defined scientific nomenclature.'

In a note to p. 1366, Nicholson and Lydekker[1] complain of the same trouble. Regarding the genera of Rhinoceroses of the American school, they say :

'From the writers' point of view the multiplication of generic terms, which, as our knowledge advances, must become less and less susceptible of exact definition, tends to drown the science in a sea of names, which form a great burden to the memory, and thus tend to destroy the very object of classification.'

Classification is not the *end* of a science, but the *means* of facilitating the conception of creation by the method of evolution ; and if the whole conception be obscured under a heap of names, its object will be surely defeated.

'La haute science' would appear to consist now in the faculty of inventing such names as the following :—'ids,' 'idants,' 'idio-plasm,' 'somatic idio-plasm,' 'morpho-plasm,' 'apical-plasm,' as composing the 'sphere' of germ-plasm, and which the late Professor Romanes[2] compared to the nine circles of Dante's *Inferno*!

I ask again, if scientists are groaning under the grip of this 'demon' of chaotic modern nomenclature, what should the poor beginner say, who would have to commit to memory such an amount of useless terms before he can understand what the professor is talking about. All this needless multiplication of terms is worrying and distressing to the 'grey matter' of the brain

---

[1] *Manual of Palæontology.* [2] *Examination of Weismannism*, p. 118.

of both professor and student, and to all earnest investigators, who wish to get to the bottom of creation by the method of evolution. As to the general reader, he may probably say: 'Non ragioniam di lor, guarda e passa.'

This troublesome multiplicity of useless words has certainly become a formidable difficulty to those who may wish to pursue scientific investigations, and an obstruction to the progress of Science; for if, before attempting to devote one's time to the study of the 'Philosophy of the Sciences,' one has to learn a language as difficult as that of cuneiform inscriptions, it will deter many from embarking in such a pursuit.

Is it any wonder that ordinary people do not think that scientific men are either so sensible and unselfish as they may think themselves to be? The curious part is that tyros may perhaps think that these strange and unpronounceable words are *the* science, and may startle their friends with the extent of scientific knowledge they have acquired at the schools, colleges, or universities!

Are then the facts of the universe, and the discoveries made by scientific explorers, to remain the possession of the few, by being locked up in a language which only means 'hieroglyphics' to most ordinary men and women?

It is truly touching to contemplate the helplessness of the human mind in face of the prodigious number of variations it has to deal with in studying organic forms.

Mr. Stebbing, in the before-mentioned work, p. 43, says: 'It may here be mentioned that the full number of joints for a malacostracean trunk leg is seven. The afflicted naturalist has for many years had to deal with these seven under the following names, coxa, basis, ischium, merus, carpus, propodus, dactylus, which respectively signify hip, foot, socket of thigh joint, wrist, forefoot, and finger or toe.

'Originally the names were longer, all being *podites*, from coxopodite to dactylopodite; to the use of these the philosophic French still adhere, though the time-saving Anglo-Saxon has for the most part rejected them ! . . . The more reasonable plan is now to denote them by means of figures from first joint to seventh joint.'

As the antennæ of the Lobster are homologous with *podites*, it is a wonder that a hundred names had not been invented to designate their hundred or more distinct joints. Why not have had also a separate name for each hair on a man's head?!

The followers of Galileo have had their revenge by pointing out the innumerable absurdities of the teachings of the Church; but the turn of the Church may come, and it may have *its* revenge!

The mischief of all this is that the mass of mankind, even in the most civilised countries, both men and women, are *wholly ignorant of the simplest facts of creation*, and all these unnecessary difficulties only increase their reluctance to have anything to do with the wonders of nature.

The craving for coining new words at every turn has already landed us in a sort of mental chaos, and earnest thinkers see that no advantage can come from this bewildering multiplicity of terms towards a simplification of science. It only burdens the memory of those who may be courageous enough to follow scientific pursuits, without in the least making things clearer.

Mr. James Geikie[1] says: 'When I attended school, the text-books used by my teachers were about as repellent as they could be.'

And at p. 12: 'Great care, however, should be taken to avoid wearying the youthful student with strings of mere names.'

When we find professors making fun of this so-called scientific

[1] *Fragments of Earth Lore*, p. 1.

terminology, we begin to hope that the tide may turn, and that we and future generations may yet be released from the mockery which some may think a good substitute for science.

The pursuits of the specialists no doubt were a great inducement to coining new names for every cell, for every joint, for every limb of a Milliped, and so forth. This, however, is what Dr. Burdon Sanderson has said, in his inaugural address to the British Association in 1893: Specialism advances knowledge at the risk of deteriorating the man, and tends to exaggerate the importance of one set of phenomena, simply because the bearing of another set is not seen in the course of that division of labour.

Taking a philosophical view of any set of phenomena does not mean that you have never attended to detail by close inspection; but it means that you are able to withdraw your mind to a distance from the detail, and so take a broader view of the whole landscape, so as to include *more* of creation at one time, and thus obtain a comprehensive view of the relativity of the detail.

John Hunter said—'Don't think, try.' He however not only tried, but also thought. The bane of modern life seems to be that there is too much trying, and very little thinking of what may happen! We should now say—'Think, investigate, imagine, and also try.' And when you get an idea into your head follow it up. If it is worth anything, and if it has any truth in it, and *if you get rid of your mental inertia, and keep your eyes and ears open*, you will be sure sooner or later to come across evidence in support of your idea. You have also of course to read and ascertain, if you can, what others have thought and have written on the same set of subjects. These are all various ways of ascertaining the truth, and of making sure that it *is* truth you are dealing with.

We have been for centuries 'hag ridden' by monstrous fictions, and it is certainly a comfort to emerge from the pressure of this

INTRODUCTION

deep sea of unrealities, and take a look round upon the upper world of realities.

At the back of the phenomenon we call a Horse, a Cat, or any other animal, there is a whole chain of phenomena—its evolution —which in ancient time was not suspected. All this chain of phenomena, leading up to what you *actually see*, has to be discovered by the aid of the imagination, which does not *always* tell the truth.

Without a free use of the imagination a dog is a dog, a cat is a cat, a cloud is a cloud, and nothing more. They are facts, like so many soldiers scattered on a field of battle without discipline or organisation. The function of the imagination is to group these scattered units into companies, regiments, and armies, and fight imaginary battles with them, all manœuvred by a general called 'Logic,' who has his eyes open, and insight to discover what is going on around him. What this general has to be particularly careful about is, not to let his imagination wander loosely, and see all sorts of things that are not justified by *ascertained facts*.

Some persons pride themselves upon not possessing any power of imagination, as if it were so very *meritorious* a feature of their 'grey matter.' They say they deal exclusively with *facts*. They do not, however, see that through this deficiency they lose that insight which is the work of what we call the imagination, and so they fail to notice what is *behind* the facts. They may perhaps not be aware that a great deal of the charm of life consists in possessing a vivid imagination, provided the possessor of it is able to keep it under control. By this faculty we are enabled, in a way, to picture what would otherwise be a wholly invisible past world. To exercise the imaginative power is to cultivate a most useful implement of research.

There is so much to learn in one's short life. Every branch of

Science is being studied with such minuteness that the help of the imagination is largely needed, not only to understand the phenomena yourself, but to make them clear to the imagination of other people. But note this: the grey matter of the human brain is a very treacherous informer. It invents a lot of things which it preaches unhesitatingly as *truths*, and then, after-generations have the task of sifting the whole, and re-classifying the supposed phenomena into truths, sub-truths, and lies!

In the course of these pages I have mentioned that the coloration of the skin of animals may have been greatly influenced by the electricity of the brain-cells. You might naturally ask—What has electricity to do with the colouring of animals? But just think of it, and tell me *where* electricity does *not* come in. The modern view of electricity is that you cannot touch anything, you cannot move anything, you yourself cannot move, or cannot think, or will, without the evolution of electricity. If I blow on my hand, the impression on my skin is electrical, and is conveyed through my nerves to my brain, and there either develops thought, or both thought and action.

One should have heard Professor Oliver Lodge on the evening of the 1st June 1894,[1] at the Royal Institution, and have seen him make experiments to illustrate the Hertz waves, in order to realise how completely the nerve-centres of animals are in the grip of their surroundings, taking of course the visceral impressions as part of the surroundings. Just as one Leyden jar in action influences another wholly disconnected with it, except through the means of the ether, but attuned to it, and sets it in motion; just as one tuning fork in vibration sets another in action which is in unison with it; just as one magnet influences another near it, so everything—light, heat, magnetism, electricity, gravity, etc., act on

[1] Lecture printed in *Nature* of 7th June 1894, p. 133.

the sensitive nerve-matter of the nervous system of animals, and influence thought and all other nerve-action.

Several branches of animals evolved from other animals which were not stationary, but were changing. While changing, they in turn were evolving others in various 'grooves' of evolution. This would account for the fact that although their descendants have several characters in common with the ancestral stock, they are nevertheless widely distinct in *other* characters; and the characters they mostly differed in were exactly those which depended on the influences of surroundings for their development, and therefore were greatly modified by them.

There is another point about which zoologists seem to have *no doubt*.

Mr. Herbert Spencer[1] says: 'Zoologists are agreed that the Whale has been evolved from a mammal which took to aquatic habits, and that its disused hind-limbs have gradually disappeared.'

Many others have also the same belief; for Mr. Hutchinson[2] says: 'Take for example the case of Whales and fishes; the original land mammal from which Whales are descended has in course of time become so fish-like in appearance that even in these modern days there are some who yet speak of them as fishes! The shape of the Whale is fish-like; it has lost its hind-limbs through disuse: it has changed its fore-limbs into paddles, which have a certain fin-like aspect; and its cousin, the Porpoise, has developed a big triangular fin on the back.'

All this derivation of water mammals from *land* mammals alone may be true or *untrue*. In the words of Mr. Hutchinson: 'What right has any one, however great his knowledge or his ability, to dictate to Nature, and to say this or that is impossible?'

---

[1] 'Rejoinder to Professor Weismann,' *Contemporary Review*, December 1893, p. 909.
[2] *Creatures of Other Days*, p. 131.

I confess that this theory which considers the fish-like mammals as having descended from land mammals which took to aquatic habits does not seem to me satisfactory. Zoologists would appear to have conceived a roundabout way of evolving a Whale. The fish is made first to evolve a land mammal, and then this takes to the water again and gets rid of its hind-legs. It seems clear to me that if the fish proper could evolve a *land* mammal, it could also evolve a *water* mammal, without the necessity of going through this roundabout performance. If a bird could lose its fore-limbs on land, it would appear that a Whale could lose its hind-limbs in the water.

This notion presupposes the possibility of *land* mammals evolving from fishes, and the impossibility of *water* mammals evolving from the same fishes. And all this in face of all we know about the Ichthyosaurs and Plesiosaurs, with their dwindling hind-limbs; in face of the shore and land fishes hopping about on their pectoral fins on land, so that they are difficult to catch; in face of the fact that certain fishes proper can breathe either by gills or by lungs, according to circumstances; and in face of the fact, which every one knows, that the Tadpole is first fish-like, and then evolves arms and legs *without getting out of the water.*

Are we so sure, in spite of 'agreement among zoologists,' that the Whales are degenerated *land* animals, and that land animals are *not* further evolutions of fish-like animals which have taken to life on dry land, while the Whale, evolving from the same fish-plane, *remained* a water mammal?

In the amphibians alone we have ample evidence that a fish-like vertebrate can grow arms and legs *without* leaving the water; and in the Ichthyosaurs we have again ample evidence that the hind-limbs were already undergoing degradation, and that in the Plesiosaurs both the hand and foot had become degraded to five digits, instead of the many digits of the Ichthyosaur form. I think no one has ever credited these extinct animals with having been first land animals which took to a water life. There is, moreover, some evidence in favour of considering them *mammals*.

Professors and authors would seem to have stereotyped on our brain the words 'normality,' 'anomaly,' 'monstrosity,' giving them certain arbitrary meanings. They perhaps may have thought that they had settled all matters regarding creation, as far as such phenomena were concerned. But let us imagine that anomalies and monstrosities may possibly have been 'factors in the origin of species.' Then we begin to see that the method of creation will appear under a somewhat different aspect from what books and professors have taught us.

Under the heading of 'Monstrosities,' I have discussed these particular phenomena, and have endeavoured to show that what we call monstrosities may have been more frequent factors in modifying the structure of animals than has been supposed.

As the whole arm can be suppressed in one birth, so, I imagine, could the Archeopteryx have had its long tail shortened to the little stump of the modern bird in *one* birth. The objection to such a sudden transformation seems to rest only in the minds of those who have worked up into an *unalterable* dogma the notion that modifications in organisms are brought about *solely* by *slow* degrees. This dogma may possess no unalterability *outside* the brain of scientists.

It might be said, if this were so, the long tail of the Archeopteryx would have revealed itself sometime by a sudden reversion. But reversions in some organisms either rarely happen, or do not happen at all, and if they do happen, sometimes they may escape notice.

For instance, no botanist doubts that the leaf of the orange and lemon trees is the middle leaflet of some ancestral form with *three* leaflets, like that of *Citrus Trifoliata*,[1] yet the one leaflet persists through millions of generations without reversions. In India, in my researches on the oranges and lemons, I sowed seeds of all kinds of Citrus, and only a *few* seedlings gave any indication in their *first* leaves of having descended from a trifoliate ancestor.[2]

Evolutionists do not believe that modern birds evolved from the extinct Pterodactyle form. Yet what is more easy than for a pterodactyle wing-membrane to grow *hair*, like other parts of the body; and for this to become exaggerated into feather-like hairs, and so on; then for the wing-membrane to contract, perhaps even in one generation, so as to envelop the arm and finger-bones, while the feathers increased in size, and became the *real* flying apparatus?

All this seems certainly preposterous and fanciful to a person who may have looked upon monstrosities as ungodly phenomena, but as they do occur now, there is no reason to suppose they did not occur in past ages.

The animal congenitally is given a certain bodily structure whether normal or abnormal.

He, that is, his nerve-centres, must make the best of it, if he is to live at all. He has, moreover, to regulate his actions and habits by the growing structure of which he is possessed, until they become established by the completion of that growth in adult age. He is the sport of inheritance and surroundings. Inheritance tends to keep him on certain more or less fixed lines, but it does not at all follow that circumstances may not change his structure to a *large* extent in one birth, so as to shunt his descendants on to a new line.

We know that all mammals are allied, for if they were not,

---

[1] See *Gardeners' Chronicle* of 18th November 1893, p. 625.
[2] See *Oranges and Lemons of India and Ceylon*, pl. 246.

they would not have the same plan of skeleton, and the same apparatus for nursing their young.

Besides nervous, circulatory, and other structural features, which mammals have also in common, there are dermic features which, I think, have not been hitherto sufficiently recognised by biologists as indications of derivation. I mean the markings on the exterior of mammals. Of course every one knows that Leopards and Spotted Cats are allied, but probably few suspect that the spots on Leopards may indicate a distinct derivation from animals which have no spots. I don't mean with Lions and Pumas and other individuals of the Cat tribe, but with animals wholly distinct from these, and even with certain extinct animals.

In works of comparative anatomy, professors show that the internal structures of mammals—bone for bone, muscle for muscle—are identical. Then whence comes all this difference of external surface? How comes it that the Leopard is rosetted, the Cheetah spotted, the Tiger and Zebra striped in one direction, while the Ocelot is striped in another direction, and so forth? Evolutionists declare that these external colours and markings have been brought about by adaptation to surroundings. In the following pages I have discussed what modification of this theory is, in my judgment, needed in order to make it conformable to all the facts that I shall place before the reader.

When we first begin to study the spotting and striping of animals, they seem a chaos of markings, unregulated by any laws; but by degrees we become aware that there is some method in the whole phenomenon, and the markings of one animal can be seen to be derived from those of another, just as in the skeleton we see each bone to be derived from that of an ancestor. If not all, most of the spotted and striped mammals, more especially the carnivora, are reducible more or less to one plan of origin.

The bibliography of spots and stripes is not very abundant, and biologists may perhaps have depended a little too much on natural selection, as being sufficient to explain the creation of everything.

The scope of some of the following studies is to show that the spotting and striping of mammals, in their origin, are not purely the results of natural selection from beginning to end. I believe them to have been originally *inherited* features, coming from very remote ancestors, and altered in many ways by transmission from species to species, from genus to genus.

So many persons are interested in Horses that to understand in some way the origin of their curious markings would add to the interest of these animals. Similarly it would add to the interest we take in our domestic animals, if we could satisfactorily account for the spotting and striping of our Cats and Dogs, and other mammals. Our surroundings would again become peopled with the remote and extinct ancestors from which those we now see have descended.

The evolutionist with his 'eyes open,' can find interest in Leopard skins, in the dapples of Horses, and in the markings of other animals; in the coloration of the legs of Horses, Dogs, Cats, and in a hundred other things which the non-evolutionist would pass by as ordinary insignificant phenomena, and totally void of interest. Fifty lives would not suffice for the evolutionist to exhaust the interest of things he may see around him. It is all a study of the real method of creation.

It is usual for people to think of the Horse as an animal fit for draught, for riding, hunting, racing, etc., but the evolutionist sees both in his internal structure and in his external coloration, relationships to animals which ordinary people think have nothing to do with the Horse.

We shall see that rosetted animals must have been *legions*

in past ages, and of all kinds and descriptions. The coloration of the skin is not a thing that can be fossilised, and so one has to put 'two and two together' in order to discover an explanation for the varied markings of the mammals of our day. We not infrequently pass over the *seeming*, and go a-hunting after the obscure and the unlikely.

It seems to me that the organs of animals which receive and store up impressions, which we call nerve-centres, are as much engaged in influencing the form and coloration of the markings of the skin as they are in moulding and modifying the skeleton and other parts of the body. They are the controllers and regulators of the whole life, not only of the individual but also of the race. And the individuals forming a race were after all part and parcel of the ancestral stock, and were at one time or other organically connected with it.

I do not, however, pretend in these pages to account for every speck and coloured hair, but to give my view of what seems to have been the genesis of spots and stripes in mammals, and of the contrasted coloration we see in so many animals, which would indicate some sort of *plan* of coloration. I do not enter into the microscopy of the subject—into how pigment cells behave in fishes, and other small animals which change their coloration and spotting according to surroundings. We know that the Leopard, the Tiger, the Zebra, and others do not do so, and therefore we have to account for the genesis of their more or less permanent spots or stripes. My suggestion would appear to be a 'vera causa' of the markings of the Jaguar, the markings of all other mammals, in cases where these exist, being only a modification of such rosettes as those of the Jaguar, and *its* markings only a modification of vastly more ancient conditions.

One of the problems to be solved is—How came the rosettes on the Jaguar, the stripes on the Tiger and Zebra, the dapples on the Horse into being? Undoubtedly they must have some reference to ancestral features. What ancestor or ancestors have these existing features been inherited from, for assuredly they present evidence of inheritance as much as the bones of their skeletons?

We should make a distinction between the general coloration of an animal and its spot or stripe colouring; both are liable to vary *independently*. The Cheetah, the Dalmatian Dog, and certain Horses and other animals are black-spotted; while certain Deer, Phalangers, and certain Horses are white spotted.

Fossils certainly give us the structure of extinct animals, but I hope to show that they can also tell us something, if not so certainly, about the probable origin of certain markings we see on existing animals. But in order to see all this a good deal of the imaginative faculty will have to be brought into play.

Probably zoologists may look upon the markings of animals as trivial and unimportant features, yet it would seem that markings, if not the general coloration, are important zoological features, and may tell a tale as interesting as that told by the teeth. Of course skin coloration and markings are more liable to change, because they have to adapt themselves perhaps more intimately to the surroundings in which the animals happen to move.

Great importance among zoologists seems to be placed on the character of the teeth of animals in grouping them for the purposes of classification, as if these were absolutely the *only* characters that are inherited.

The reason why so much importance has been given to teeth as a character indicating descent is that fossil vertebrates have rarely anything but their simple skeleton to show what they may have been like, and certainly the teeth may indicate their habits. The skin characters have usually wholly disappeared, and we have nothing to guide us, in that direction, but the skins of existing animals.

There is, however, some evidence which would tend to make us suspect that teeth may be liable to sudden changes, owing to contraction of the jaws. Teeth, like other bones, it would appear, are subject to fusion or to dissociation, as the case may be. And the writer on Seals in the *Royal Natural History* quotes an interesting example of dissociation in teeth, which I have quoted more fully in another place, as I think it very instructive, and the inference to to be drawn from it important.

The doctrine of evolution has replaced every other doctrine of creation, owing to the undeniable support that existing facts give it. This being unstintedly admitted by all modern scientists, and by many modern theologians also, there remains only to account, in some way, for the appearance on this earth of the innumerable creatures we see, including man himself, by the method of evolution.

Lower down in the scale of life, beyond a certain stage, we cannot go, in this investigation, because breaks occur which are at present in no way filled up. Whether the gaps may or may not be filled up at some future period no one living can say.

The evolution of the structure of one kind of animal from another has been made clear enough, but the evolution of the coloration of one kind of animal from another has not been made sufficiently clear. Probably this feature in evolution has been neglected, because it offered difficulties which perhaps looked like *puzzles*. It is this feature of evolution which I have tried to make clear in some of the following pages. With regard to drawing conclusions, Professor Huxley[1] remarks :—

'What in fact lay at the foundation of all Zadig's arguments

[1] 'On the method of Zadig,' *Science and Hebrew Tradition*, pp. 7 and 8.

but the coarse commonplace assumption, upon which every act of our daily lives is based, that we may conclude from an effect to the pre-existence of a cause competent to produce that effect.

'Zadig was able to do this because he perceived endless minute differences (and likenesses) where untrained eyes discern nothing ; and because the unconscious logic of common sense compelled him to account for these effects by the causes he knew to be competent to produce them.'

SPOTTED AND STRIPED MAMMALS
(HORSES EXCEPTED)

'YET, if he would be guided by the true spirit of scientific inquiry, he must maintain an unsettled opinion as long as the evidence is incomplete or contradictory; he must adopt conclusions only where the evidence is complete and convincing; he must ever hold his mind open to new evidence, even if it bring about the abandonment of accepted beliefs. He may, if desirable, quote the conclusions of others, and, if well read, he may thus become widely informed; but he will fail to gain the best benefit that comes from careful study, if he does not reach opinions and conclusions for himself, forming them only as fast as the evidence that may support them is clearly understood.'

Elementary Meteorology, by Prof. W. M. DAVIS.

PART I

SPOTTED AND STRIPED MAMMALS

(HORSES EXCEPTED)

A GLANCE at the living Mammals in the gardens of the Zoological Society, and at the mounted specimens in the Natural History Museum, will show us what a large number of Mammals of several orders and of many genera are either spotted or striped, or both spotted and striped. There is a large number of Mammals which may be said to have permanently lost their spots or stripes; but there are not very many which, either in the childhood of the individual or in some of its species, either on the legs, on the tail, or on other parts, do not betray their descent by vestiges of either spots or stripes.

In the Appendices I have given lists of various Mammals which show spotting or striping now, or show vestiges of descent from spotted and striped ancestors. Some of them have a plain body and spotted or striped legs; while others have only a ringed tail to show what they came from. These tail rings, even when they are the sole markings, are in my opinion distinct vestiges of either a spotted or striped ancestry.

If one had a fuller acquaintance with the childhood and adulthood of all Mammals, under different conditions of climate and other surroundings, the probability is that these lists might be much lengthened, and we might perhaps then come to the conclusion that most Mammals, at least, had a spotted ancestry more or less remote, not even excluding the Marsupials of Australia.

The earliest record that I have met with of a striped Mammal is that shown in Fig. 1. It is the bone handle of a poignard of a prehistoric period. It represents some kind of deer which had partial broad stripes on its flanks, evidently vestiges of something like the zebra bands in the same regions.

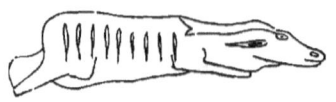

FIG. 1.—Cast of a handle of poignard found at Bruniquel, on the river Aseymn, France. (Brit. Mus.: Mammoth and Reindeer period.)

It might perhaps be thought, as an alternative, that these marks were not intended by the carver to indicate skin-stripes, but merely the projections of the ribs.

If, however, we consider that, if the prehistoric savage knew anything, he must have known a *great deal* about the ribs of the animals he was continually hunting, cutting up, and feeding upon, we shall see that he must have known that the ribs of the deer did not extend to its haunches. Therefore, the transverse stripings on this ancient model of a deer can hardly be taken to have been meant to indicate the projections of the ribs, but are more likely to have been meant for skin-stripes. Moreover, if the reader will turn to Appendix A, Fig. 22, he will see an antelope with striping not very unlike that of this prehistoric relic.

I shall, however, leave this point to be decided by archæologists and palæontologists, and proceed with my story.

What causes the changes in the markings of different animals, and at different ages, I do not know. Presumably atomic changes in the nerve centres, initiated by surroundings, heredity, or what

SPOTTED AND STRIPED MAMMALS 5

not, become reflected electrically on the skin, whether during the embryonic stage or afterwards, and cause aggregation or dissociation or other changes in pigment cells.

Evolutionary biologists—and probably there are at present few or no biologists who have *not* accepted the doctrine of Evolution—seem inclined to consider that these markings in animals are the result of natural selection, acting cumulatively on some fortuitous variations that may have occurred, and do occur, in an infinity of ways. By natural selection is meant the weeding-out, generation after generation, of all those variations which are insufficiently protected by their surroundings for either offence or defence, or both, and by keeping alive those which are most fit. Reproduction and heredity then maintain and improve this selection.

Dr. Alfred Russel Wallace says[1] that 'Professor William H. Bremer of Yale College has shown that the white marks or the spots of domesticated animals are rarely symmetrical, but have a tendency to appear more frequently on the left side. This is the case with Horses, Cattle, Dogs, and Swine. . . . Among wild animals, the Skunk varies considerably in the amount of white on the body; and this, too, was found to be usually greatest on the left side.[2] A close examination of numerous striped or spotted species, as Tigers, Jaguars, Zebras, etc., showed that the bilateral symmetry was not exact, although the general effect of the two sides was the same. This is precisely what we should expect if the symmetry is not the result of a general law of the organism, but has been, in part at least, produced and preserved for the useful purpose of recognition by

[1] Note to p. 217, *Darwinism*.
[2] I should say this is a sure indication that the difference does not depend on the *skin*, but on the *unequal* action of the two halves of the nerve centre.

the animal's fellows of the same species, and especially by the sexes and the young.'[1]

Then on p. 199, quoting from Major Walford, a Tiger-hunter, Dr. Wallace says: 'There can be no doubt whatever that the colour of both the Tiger and the Panther renders them almost invisible, especially in a strong blaze of light, when among grass; and one does not seem to notice stripes or spots till they are dead.'

I suspect the 'strong blaze of light' had something to do with the invisibility on the part of Major Walford, for he says that natives could see the Tiger, which would seem to mean that their eyes are accustomed to strong light, and can adapt themselves to it.

There cannot, however, be any doubt that the two sides of a spotted or striped animal are unsymmetrical. A glance at the Tiger and Leopard skins in the London fur-shops would be enough to convince any one of this. And it is, I think, due to a want of *identical* nervous action in *both halves* of the central nerve organ, to the atomic action of which I would attribute *all* skin colorations.

One of the objects of these pages is to investigate how far the markings of animals are due to natural selection, and how far they are not.

The innumerable variations in the markings of horses, which we see in the streets of London, will be made to contribute evidence in this interesting investigation.

In many cases it will not be difficult to show that the striped animal is only a modification of the spotted animal.

[1] The probability is that wild animals recognise their fellows more by scent than by sight; nevertheless, it is curious to note how dogs recognise dogs *of any breed*, at a distance; they, however, complete their investigation by means of the nose.

SPOTTED AND STRIPED MAMMALS

We shall first examine some of the most markedly spotted and best-known animals, viz., the Jaguar, the Leopard, the Cheetah, the Ocelot, and the Serval. For my purpose the Jaguar and Leopard, and also the Panther, may be considered as *one* animal, the others being differently marked.

Mr. G. P. Sanderson[1] says: 'The distinction between the Panther and Leopard is practically small, and lies chiefly in the inferior size of the Leopard. The markings, habits, and general appearance (except size) of the two animals are almost identical. But neither can be confounded with the *Cheetah*, even by the most casual observer . . . the spots of the Panther and Leopard are grouped in rosettes, enclosing a portion of the ground colour;[2] whereas those of the *Cheetah* are solid, and are separate from each other.'

Mr. Blandford[3] declares that there is no difference whatever between the Panther and Leopard, and Mr. Blyth was of the same opinion. He also states that black and ordinary Leopard cubs are often found in the same litter, and that an albino Leopard is figured in Buchanan Hamilton's drawings.

Mr. Sanderson further states that the black Leopards from Java have all sorts of shades, from jet-black to light brown; and that the black Leopard seems to be confined, at least in India, to heavy forest tracts, while the common variety in Mysore frequents open country, and also rocky localities.

It should be here noted that in black Leopards, as in certain black Cats, the markings are often plainly visible in certain lights. The marking is persistent, and quite independent of melanism, or that condition which produces the general blackness of the skin.

[1] *Wild Beasts of India*, p. 327.
[2] Sometimes of a *different* shade from the ground colour.
[3] *Mammals of India*, p. 68.

8 STUDIES IN THE EVOLUTION OF ANIMALS

The Jaguar is only a South American Leopard, and black variations of it are frequent also there.

It may be interesting to note that the black Leopards of Africa differ somewhat from the black Leopards of Asia.

The *Royal Natural History*, vol. i. p. 338, regarding the black African Leopards, states that in 1885 a black specimen, obtained near Grahamstown, was described by Dr. Günther. Its ground colour was a rich tawny, with an orange tinge; but the spots, instead of being of the usual rosette-like form, were nearly all small and solid, like those on the head of an ordinary Leopard.

In the black Leopards of Asia the rosettes are retained, while in those of Africa they appear to lose their ocellus and become solid. The jet-black Leopards, like the jet-black domestic Cats, usually lose all traces of markings.

Leopards and Jaguars are tree-loving animals, and therefore it seems obvious to evolutionists that their markings were the result of natural selection, acting cumulatively on favourable variations so as ultimately to harmonise them with a surrounding of speckled lights and shades produced by the leaves of trees.

Unlike the Lion and the Tiger, the Leopard of India is 'thoroughly at home in trees, running up a straight-stemmed and smooth-barked trunk with the speed and agility of a Monkey';[1] and Mr. Hunter remarks that in Africa 'the Leopard nearly always puts the remains of his "kill" up a tree.'[2] Then the Jaguar 'is one of the most expert climbers among the larger Cats';[3] and during inundations it is said that it will sometimes take to an arboreal life, preying upon Monkeys.

So we see there is ample evidence to show that the Leopards

[1] *Roy. Nat. Hist.*, vol. i. p. 390.   [2] *Ibid.* p. 392.   [3] *Ibid.* p. 395.

SPOTTED AND STRIPED MAMMALS 9

have their markings in harmony with the surroundings of an arboreal life.

FIG. 2.—Jaguar, from a photograph by Gambier Bolton, F.Z.S.

FIG. 3.—Leopard, from a photograph by Gambier Bolton, F.Z.S.

A glance at the markings of the Jaguar and Leopard in Figs. 2 and 3 will show that a large number of their rosettes is made up of *groups* of small spots, each group forming an isolated and irregular ring of small black spots with an enclosed space. This space may be either of the same colour as the general ground of the skin, or of a darker shade, and sometimes of a different hue. Moreover, on the Jaguar skin (Fig. 4) there are many rosettes which have one or more small black specks in the *middle* of the enclosed space, which in Leopards proper seem to be obliterated, owing perhaps to a contraction of the entire rosette.

In the Tring Museum there is a fine specimen of a Jaguar. On its flank are very large polygonal rosettes, with from *one to six* specks in the enclosed space.

If any one will take the trouble to look over the Leopard skins in the windows of the London furriers, he will be at once convinced that the rosettes even in the same skin vary immensely;[1] and if different skins are compared it will be found that, although the general mapping may be similar, the detail shows that there are scarcely two skins alike. Indeed, the skins are as different as the faces one sees among the people in the crowded streets of London. It seems astonishing that, among the thousands of faces one sees, there should not be two alike. It is the same with the coloration of most animals.

Then, if we examine the skins of Mammals which are supposed to be of different species, although of the same genus, we find astonishing modifications of what I consider the typical rosettes of the Jaguar.

A very interesting monograph of the *Felidæ* by D. G. Elliot shows, by means of the beautifully coloured plates, not only the modifications of rosettes, but all manner of intermediate stages up to total obliteration of all markings. The transitions from rosettes to spots and stripes can there be readily seen.

---

[1] A variety of Jaguar from Mexico is characterised by the distance at which the small spots which ordinarily constitute the rings are placed from one another, so that complete rings or rosettes of spots only occasionally occur. *Roy. Nat. Hist.*, vol. i. p. 395.

SPOTTED AND STRIPED MAMMALS

I can only give a small number of Cats in this superb monograph, to show the principal variations from the typical rosettes of the Jaguar.

In Elliot's Jaguar (*Felis onca*), which presumably was copied from nature, there are on the flanks very distinct rosettes, made up of polygonal rings of black spots, more or less fused, and enclosing a space which is differently coloured from the inter-rosette ground ; and each rosette has a distinct black speck in the centre of the enclosed space.

Then his Margay (*F. tigrina*) is of a Leopard-yellow colour, rosetted in various ways, the rosettes being made up of three, four, and five black spots, which enclose a brown space. It is, moreover, distinctly barred on the shoulder and back. (See another variant on p. 418 of *Roy. Nat. Hist.*, vol. i.)

Fontanier's Spotted Cat (*F. tristis*) is something like a Jaguar, but its rosettes are distorted in various ways.

From this we pass to his African Golden Cat (*F. chrysostrix*), which is either grey or brown trimmed with Leopard-yellow. Its spots are solid.

The Serval (*F. serval*) is much the same ; only, in addition, it has fusions of spots into longitudinal streaks.[1]

The Rubiginous Cat (*F. rubiginosa*) is of a brownish-grey, with solid black spots arranged in *longitudinal rows*, preparatory to fusing in longitudinal stripes, like those on the back of the neck.

We come then to the Pampas Cat (*F. pajeros*). It has brown longitudinal bands on grey ground, in the manner of the Ocelots, and the legs are transversely banded. (See variant of this on p. 431, *Roy. Nat. Hist.*)

The Clouded Tiger is very interesting (*F. diardi*). It has a

---

[1] The Serval is also subject to melanoid variations, and the spots are distinctly visible when viewed in certain lights. (*Roy. Nat. Hist.*, vol. i. p. 414.)

yellowish-brown general colour, with broad transverse patches of a yellowish-grey, margined with black blotches or spots. The patches are evidently fusions of several rosettes of a similar colour. The haunches are rosetted, and the tail has its rings double, which is also a vestige of a rosetted body.

There is another much like the foregoing, the little Marbled Tiger (*F. marmorata*).[1] It is either Leopard-yellow or grey, with large clouded patches edged and spotted with black, while the general colour is paler. Its haunches and shoulders are spotted. The tail is either spotted or ringed.

We come now to the Caffer Cat (*F. caffra*), which is of a bluish-grey, striped with black, Tiger-fashion. (See variant on p. 421, *Roy. Nat. Hist.*)

From this we pass to the Tigers, which every one knows.

We pass, then, to total obliteration of spots and stripes in the self-coloured Cats, like the Puma (*F. concolor*), which is also called Cougar, Panther, and American Lion. In the adult stage it is all plain, and of a rich brownish-grey, but its kittens are *spotted*.

How astonished the Puma must be when she has cubs for the first time! She looks at her husband's coat and at her own, and sees them of a uniform rich isabelline colour, and then she finds her kittens are born spotted all over like young leopards. Are these really my children? Yes, your very own! *You* have succeeded in shaking off your rosettes, but your kittens still masquerade in that antiquated dress, and prove to you that after all your pedigree is identical with that of the Leopard!

There are innumerable transition markings between rosettes, solid spots, and stripes, and many Cats have only *vestiges* of spots or stripes. The tails of most of these *Felidæ* are ringed, and the

---

[1] In the *Roy. Nat. Hist.* these are called Clouded *Leopard* and Marbled *Cat*.

under surfaces of most of them are *paler* than the back and flanks, and in some cases wholly *white*.

Elliot's monograph of the *Felidæ* contains gradations, modifications, and transformations of rosettes and spots, which can be studied with comfort within the compass of a book. It is like a museum of Cats.

To facilitate the examination and comparison of Leopard and Jaguar rosettes, and to show both flanks at one glance, I have given in Figs. 4-7 some skins spread out; and Fig. 59 gives a number of variations of single rosettes taken from numerous Leopard skins.

It will be seen that on the Jaguar skin there is a large number of rosettes consisting of an irregular or polygonal ring of small spots enclosing a space, in the middle of which, as I said, there are one or more specks. At times the ring-spots are dissociated, as on the shoulder, and they appear like an irregular group; at other times the ring-spots coalesce wholly or partially, and form a more or less continuous polygonal ring, as in those of Fig. 7, with or without the central specks. The rosettes of what are commonly called Leopards are usually *without* the enclosed specks.

This continuous ring can best be seen on the Leopard skin of Fig. 7, already alluded to.

Again, we see that on the abdominal surface the rosettes tend to coalesce further, with obliteration of the enclosed space, and, in the Jaguar, to form a sort of trefoil, quadrifoil, pentafoil, etc.

I would here note that on the tails of these Leopards the rosettes, at first isolated, tend to coalesce and form transverse rings towards the tips, with obliteration of the enclosed space; and that along the spine the rosettes tend to coalesce longitudinally, and to form a continuous dorsal line or band.

All variations of Leopard rosettes would seem to be modifica-

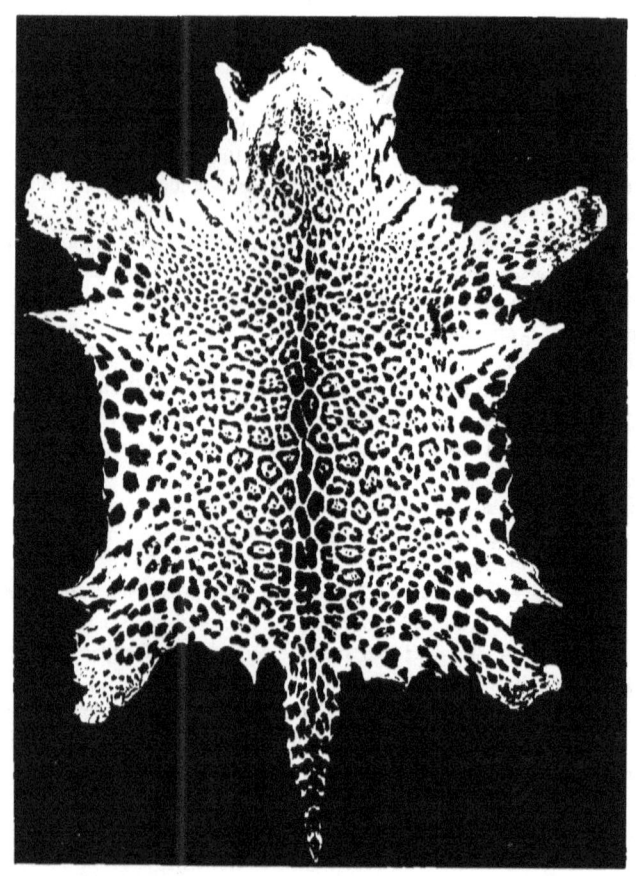

FIG. 4.—Skin of Jaguar, from a Photograph by Messrs. Dixon and Son.
Skin lent by Messrs. Jeffs and Harris.

FIG. 5.—Skin of Leopard—may be African; from a Photograph by Messrs. Dixon and Son. Skin lent by Messrs. Jeffs and Harris.

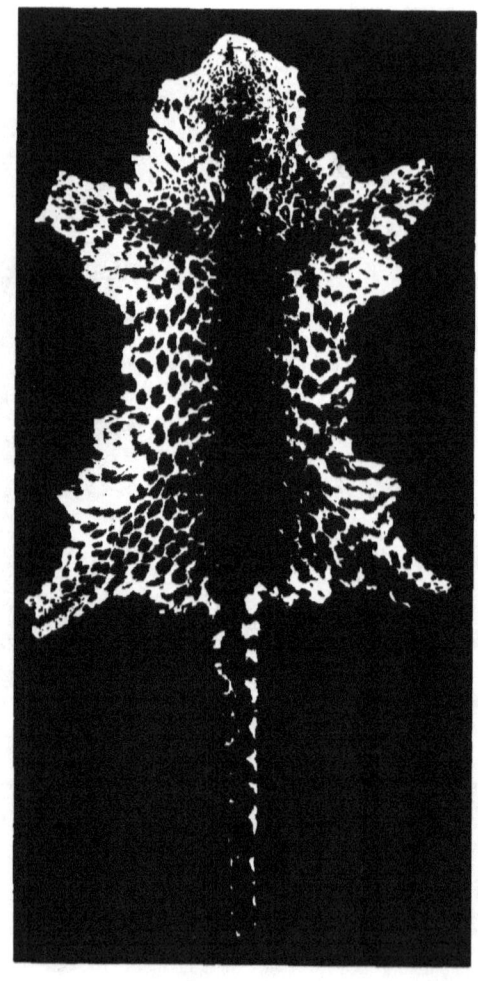

FIG. 6.—Skin of Leopard—probably Chinese; from a Photograph by Messrs. Dixon and Son. Skin lent by Messrs. Jeffs and Harris.

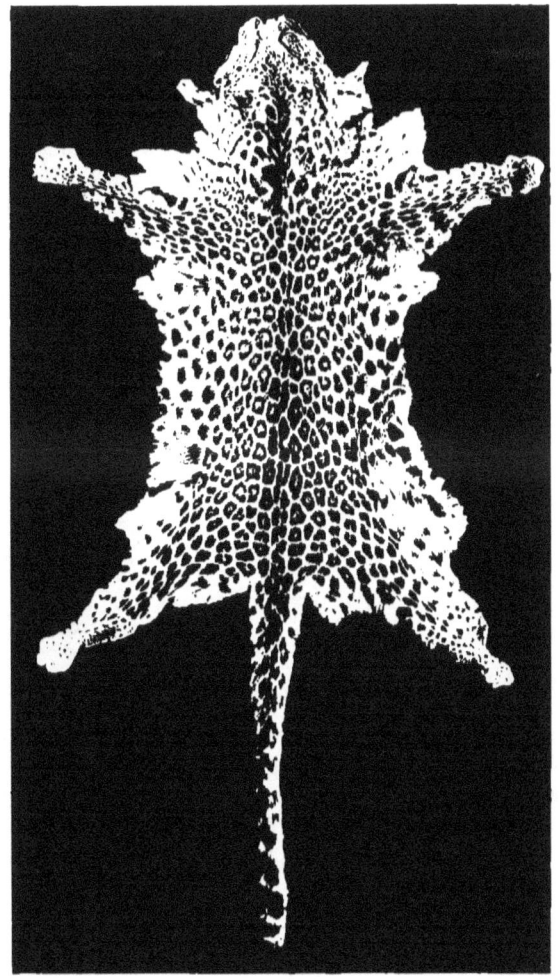

FIG. 7.—Skin of Chinese Leopard, from a photograph by Messrs. Dixon and Son. Skin lent by Messrs. Back and Co.

tions of those on the *flank* of the Jaguar; and Fig. 4 shows many intermediate forms between the Jaguar rosettes enclosing specks and the solid rosettes or spots of the abdominal region.

In one particular Leopard skin[1] I noticed a very curious variation, shown in Fig. 8. It appeared as if the enclosed specks had been *extruded* from the rosette ring.

FIG. 8.—Occasional variants of Jaguar and Leopard rosettes.

In some regions it is not always easy to make out whether the rosettes are a coalescence or a dissociation of spots.[2]

Fig. 9 shows rosettes from the scapular regions of a Jaguar skin. Some look like a consolidation and others like a dissocia-

FIG. 9.—Various forms of rosettes from the scapular regions of a Jaguar skin.

tion of spots. The groups shown in Fig. 59 (Nos. 30-32) are obviously a dissociation of the ring-spots.

The Jaguar in the Science and Art Museum, Edinburgh, has the spaces enclosed by the rosettes of the whole skin of a deeper shade of fawn than the general ground colour; and on the hind-legs

[1] Shown to me at Messrs. Back and Co.'s.
[2] Two Leopards, described by M. A. Milne Edwards, 'were remarkable for the circumstance that the markings on the flanks were more like rings than rosettes' (p. 390, *Roy. Nat. Hist.* vol. i.).

it has fusions of the ring-spots into bigger spots or blotches, with a deeper shade of fawn colour between them than the general fawn colour. I have endeavoured to show this in No. 9, Fig. 59.

Moreover, some Leopards, such as that of Fig. 6, have large solid rosettes of irregular shape on their haunches, while those on their flanks are ocellated.

This, I think, is clear evidence that, in these cases at least, the haunch rosettes are mere contractions of the larger typical rosettes on the flanks.

A glance at the Jaguar skin of Fig. 2 also shows distinctly that the *enclosed* spaces of the rosettes are of a deeper shade than the general colour *between* the rosettes. I mention these details because in these Mammals there appear to be three distinct colorations, viz., the colour of the inter-rosette spaces, of the rings, and of the enclosed space. All three may vary *independently* of the others, not only in colour, but also in form.

The different colours of the inter-rosette spaces, of the spots, and of the enclosed spaces, would seem to indicate that each has a separate and distinct nerve-centre, as much *localised* as the centres of the different parts of the arm, the leg, the face, etc., and that each of these components of the whole surface may vary *independently* of the others. It seems curious that the spots of the Dalmatian Dog should be black on a white ground, and those of the Phalanger and Dasyure white on a black ground.

The general coloration of Mammals seems of little importance, as it varies in almost every individual; what is tan in one may be either black or white in another. But the 'markings' and the colorations, which are seen to be like a sort of 'plan,' are of much greater importance, as they more or less indicate, I think, something inherited from *very remote* ancestors.

Let us now take a look at a very differently spotted Mammal.

Fig. 10 shows the picture of a living Cheetah, and Fig. 11 the skin of a similar animal (perhaps an older one) spread out to see both sides at once. In the Cheetah we find numerous solid circular spots, with minute specks interspersed among them. The large spots are disposed in transverse rows on the flanks (Fig. 10).[1] The minute specks, however, in the figure of the skin are interspersed among the larger spots, apparently without any order;

FIG. 10.—Picture of a living Cheetah, from a photograph by Ottomar Anechütz, Lissa (Posen).

while in the figure of the living animal the minuter specks appear to be disposed in many places in rows also, alternating with the rows of the bigger spots.

In the Cheetah it is not easy to make out whether the larger spots are consolidations of the *entire* Leopard rosettes, or dissociations of the spots forming the rosette *rings* of the Jaguar. The

[1] At the International Fur Stores, Regent Street, I was shown a Cheetah skin with some of the rows on the flank undergoing fusion, and forming beaded strings; and several couples of spots were actually fused into one blotch.

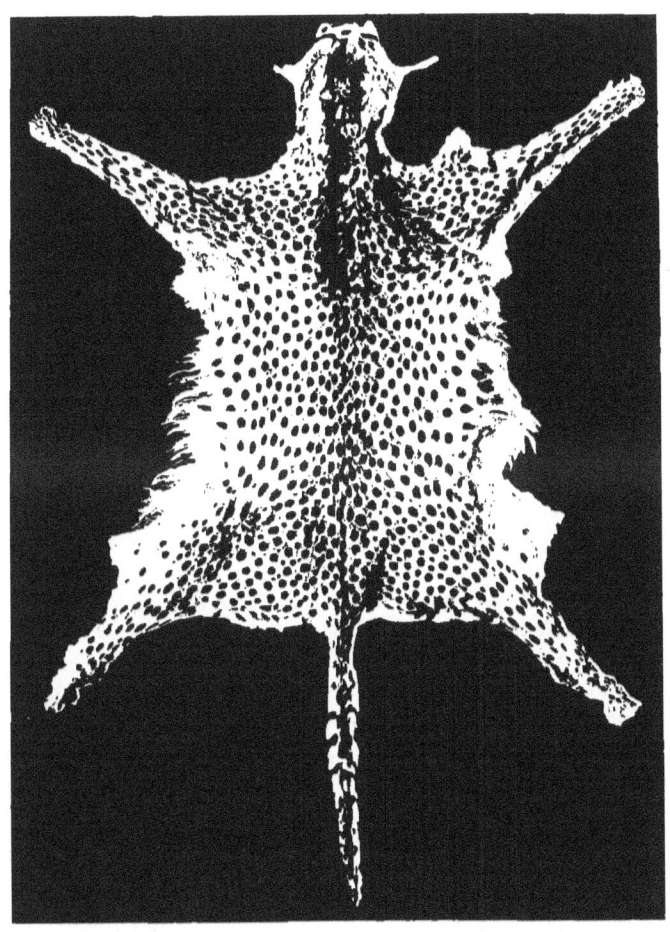

FIG. 11.—Skin of a Cheetah—perhaps of an old one; from a photograph by Messrs. Dixon and Son. Skin lent by Messrs. Back and Co.

spots on the Jaguar's shoulder (Fig. 4) are evidently *dissociated* rosettes, while the spots on its hind-quarters and abdomen are evidently *fusions* or consolidations of rosettes; so that the Cheetah spotting may have had either the one or the other origin.[1]

Anyhow, it is evident that the transverse strings of spots on the fore-legs of the Cheetah (Fig. 10) are *homologous* with similar transverse marks on the fore-legs of the Jaguar and other Leopards.

As to the minute specks, I have a suspicion that they may have possibly resulted from the specks enclosed within the Jaguar rosettes. In the modifications which these animals have undergone, the specks may have been extruded, as we have almost seen them do in Fig. 8, and have become disseminated among the bigger spots.

A close scrutiny of the Cheetah spots may lead one to detect something like dissociated rosettes, especially on the right shoulder and haunches of the skin figure; but on the haunches and tail of the living-animal figure the spots look more like *consolidations* of whole rosettes.

There is no good reason why the characteristic spotting of the Cheetah should not be a combination of both processes, viz., dissociation in some parts and consolidation in others; for in the *same* animal—the Jaguar—we find typical rosettes on the flanks, dissociated rosettes on the shoulders, and consolidated rosettes on the abdomen and legs. In certain Leopards the enclosed specks become entirely obliterated, while in the Cheetah these little specks may form one of its characteristic features.

Fig. 59 (No. 35) shows four groups from the flank of a Cheetah; and on the haunch of a Cheetah in the Natural History Museum,

---

[1] Of two Cheetahs in the Tring Museum, one has isolated spots on half its tail, while the other has, in the same part, *rings* with scolloped edges, indicating a *fusion* of spots.

SPOTTED AND STRIPED MAMMALS

as shown in Fig. 12, there is a similar disposition of large round spots and minute specks. From this one might perhaps surmise that the circular spots in the Cheetah represent the modified *enclosed space* of the Jaguar rosettes, while the specks represent the dissociated *ring of spotlets*. The change of colour of the enclosed space from brown to black is not wholly imaginary, for in horses we find that the white spots of dappled animals are changed to black spots in the roan and to brown spots in the strawberry roan.[1] I confess, however, that the Cheetah spotting is rather puzzling, for the individual spots are as round as a shilling, with a general equality of size, and they do not give any indication of a coalescence of minor spots like those on the abdomen of the Jaguar. Yet the consolidated rosettes on the paws of the latter animal are not unlike those of the Cheetah. They have, moreover, minute specks among them. As spots can be wholly obliterated, so, I suppose, they can diminish in size.

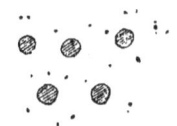

FIG. 12.—Group of Spots from the haunch of a Cheetah (Nat. Hist. Mus.).

In a Leopard skin there were here and there minute specks, like those of the Cheetah, interspersed among the usual rosettes. This is uncommon in Leopard skins; but in the Natural History Museum (case 12) a Leopard among the consolidated rosettes of its fore-legs has a number of minute specks like those of the Cheetah; and in the window of the International Fur Stores I saw a Leopard skin which had small and roundish spots on the back (shoulder region) which were not unlike the larger spots of the Cheetah. So that it is not improbable that the larger circular spots of the Cheetah may

[1] We may have melanism or albinism or shades thereof on the *entire* surface, or changes of colour in *certain parts* only, as I shall endeavour to show in another place. In another place I have also indicated that black, tan, and white are interchangeable colours.

24   STUDIES IN THE EVOLUTION OF ANIMALS

be, after all, *consolidated* Leopard rosettes further modified into circular spots.

On the other hand, Fig. 12 shows conditions which would suggest that it is the *enclosed* spaces of the Jaguar rosettes which become the large round spots of the Cheetah, while the specks are the dissociated rings of spotlets; only the enclosed space in this case would be, as I said, changed from irregular brown to circular black.

A very young Cheetah in the Natural History Museum (case 16 —*Gueparda jubata*), from the Cape of Good Hope, shows a very interesting variation. It is brown with faint spots, but what is still more curious is the fact that its back is *grey* like that of a badger!

It may be of some interest to show how much variation the Jaguar rosettes can undergo. Fig. 13 is taken from a Jaguar pictured in Griffith's *Cuvier*. They may, perhaps, be closely matched from groups on the Cheetah skin of Fig. 11.

FIG. 13.—Rosettes from a picture of the Jaguar in Griffith's *Cuvier*, p. 455, vol. ii.

This much is clear to me, that the Cheetah and the Leopard are closely allied in habits and structure, and their spotting, however modified it may have become, must have had *one* ancestral origin, not necessarily of course from the same *individual*, but from the same *species* of ancestor; and that the difference in the existing animals comes from microscopical changes in the nerve-centres, which would result in pronounced differences on the *skin*.

The student of animal markings would do well to study, as a previous training, the many-synonymed orchid—*Odontoglossum crispum*, and others of the same genus. Nothing is more interesting than a review of the variations of blotches on the petals of this genus. There are blotches of various sizes and forms; there are cross-bars; in some varieties there are *single* little spots on each petal; and, finally, in others every trace of blotching or spotting disappears, and the whole flower is white, with a little lemon-colour on the lip. In the *Dictionary of Gardening* it is stated that 'it is a plant which varies to an almost endless extent, no two of the many thousands imported being perhaps exactly alike.'

FIG. 14.—(*a*) Young and (*b*) Adult of Spotted Deer (*Axis maculatus*, Nat. Hist. Mus.).

Not impossibly, also, the stretching of the skin as the animal grows may, in some instances, tend to modify the grouping of the spots, and have something to do with dissociation. On the other hand, contraction of the skin in other regions may have something to do with consolidation of rosettes.

Perhaps, in illustration of the former conditions, one might take the case of the Spotted Deer shown in outline in Fig. 14. The young one (*a*) has its white spots close to each other, in some parts almost coalescing into blotches. Then as the animal grows the spots become more distinct, and in the adult they are separated by long intervals, as shown in (*b*).

However this may be, we cannot ignore the fact that the whole physiology of the skin in Mammals, with its colourings and markings, is under the control of the nerve-centres (perhaps as much as electro-plating is under the control of the battery or dynamo), and these again are under the various influences of heredity, age, variations of temperature, climate, composition of the blood, and other surroundings, of which we know yet too little.[1]

In the various kinds of Leopards we see that not only the rosettes differ, but also the spaces between them differ much.

These inter-rosette spaces run into each other, and form a sort of broad reticulation which is the ground-colouring of the skin.

The Leopard of Kismaya, British East Africa,[2] has its rosettes much closer than those of the Indian Leopard or of the American Jaguar, so that it seems much darker than they. The comparative smallness of this animal, supposing the number of rosettes to be equal, may, I think, sometimes account for the compactness of the rosettes, as well as for their elementary spotlets.

At Mr. Rowland Ward's establishment, Piccadilly, I was shown a very small fœtal Leopard, that is, in a stage *before* birth. It was closely spotted all over, but none of the spots were ocellated. Another young Leopard, but older than the preceding, had a commencing faint ocellus in some of its spots. On the other hand, a very

[1] Orchids seem to undergo similar alteration in the spotting of their flowers, without any nervous influence that can be detected. Yet there must be some means of communication in the *Dionæa* between the bristles and the hinges of the leaf-blades, analogous to nervous or electric communication.

[2] In the Zoological Society's Gardens.

young animal in the Natural History Museum (case 13), ticketed as a young Jaguar, has *no spots* at all, but is of a *uniform brown*.

The variations in the disposition of the rosettes of Leopards are very considerable. In some specimens they are distributed irregularly, in others they occur in slanting rows, as in Fig. 15 (*a*). Where they are crowded, two or more fuse into an elongated ocellus, as in (*b*). The Fig. 59 shows how numerous are the variations in individual rosettes.

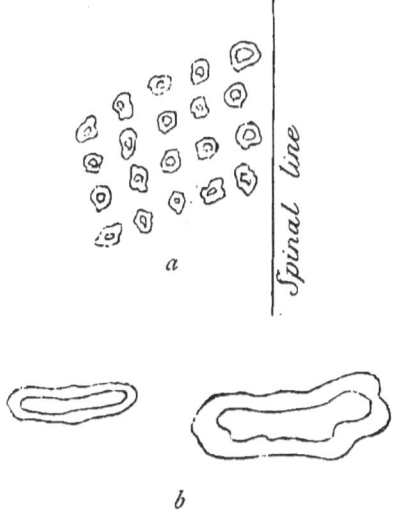

FIG. 15—(Diagrammatic disposition of leopard markings, both taken roughly from skins in furriers' windows. (*a*) Rows of rosettes, which might fuse into stripes. (*b*) Fusion of two or more rosettes.

We should make a distinction between the general colour and the spot or rosette colour; both, as I said, are liable to vary *independently*. The Cheetah and the Dalmatian Dog are black-spotted, while the Deer is white-spotted. The ordinary Leopard has a general tan colour, the melanoid a general brown colour, and the Snow-leopard[1] a general white colour, although the rosettes remain black in all cases.

[1] Of two Snow-leopards in the Tring Museum, one has ocellated rosettes, and the other has a large number of the rosettes *solid*, especially on the shoulders, haunches, and lower part of flanks.

In these variations we again find a parallel in *Odontoglossum*. Some species are pure *white*, with maroon blotches, or spots, or bars, while others are *yellow*, with maroon blotching.

The spots on the legs and tail of the Lion of Fig. 16 (*a*) leave no doubt that the ancestors of the Lion and Leopard were *one*. In adult Lions the rosettes become more and more obliterated, but in the young they cannot be mistaken for other than Leopard-spotting. The same may be said of the spots on the fore-leg of the Puma shown in Fig. 16 (*b*). In the Science and Art Museum of Edinburgh there is a largish Cat, ticketed *Puma*, which may be a young one. It is of a light reddish-fawn colour,[1] with distinct spots all over it of a deeper fawn. Its present general colour is not unlike that of the red domestic Cat.

The *Encyclopædia Britannica* says: 'The young of the Puma, as in the case with the other plain-coloured *Felidæ*, are, when born, spotted with dusky brown, and the tail ringed. These markings gradually fade, and quite disappear before the animal becomes full-grown.'

Some varieties of Lynx, although their backs are plain, have spots on their abdomen and legs. These may be seen in the Natural History Museum.

Then there is a large number of widely different animals, such as Racoons, Lemurs, and many others,[2] which, although they have neither spots nor stripes on their bodies and legs, yet have distinct rings on their tails, like those of the Leopards and Tigers of our illustrations. Therefore, all these ring-tailed animals should, I think, be credited with either a spotted or a striped *ancestry*.

In the Appendix I have given a list of animals, of very varied natures, which have ring-tails. They probably *all* descended from spotted ancestors, and the marks on their tails are the *only vestige* which now indicates the history of their ancestry.

---

[1] Skins become faded in time.   [2] See Appendix E.

(a)

(b)

FIG. 16.—(a) Lion, and (b) Puma, from photographs by Ottomar Anschütz.

30 STUDIES IN THE EVOLUTION OF ANIMALS

As to the markings of the Serval in Fig. 17, it is not likely that any one will take them for any other than consolidated Leopard rosettes placed widely apart, and in places arranged in longitudinal series.[1]

FIG. 17.—Serval, from a photograph by Ottomar Anschütz.

In Fig. 18 is given a Marbled Cat, which, although ancestrally rosetted, has its spotting undergoing obliteration, like the adult Pumas and Lions.

We now turn to the numerous variations in the markings of Ocelots.

[1] Note the black mark from the heel to the toes of hind-legs. It is an interesting feature, to which I shall refer in another place.

FIG. 18.—Marbled Cat : on the flank it has distinct rosettes, and on the legs transverse stripes. From a photograph by Ottomar Anschütz.

32 STUDIES IN THE EVOLUTION OF ANIMALS

Fig. 19 (a) shows one variety in which the typical Leopard rosettes are plainly recognisable, only they are arranged in longi-

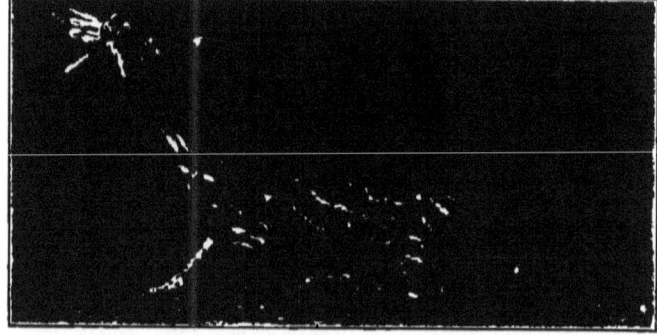

FIG. 19.—Two distinct variations of Ocelots, from photographs by Ottomar Anechütz.

tudinal order, and on the flanks and shoulder the rosettes coalesce into parallel bands, with the enclosed spaces also continuous. It

SPOTTED AND STRIPED MAMMALS 33

should be noted that on the shoulder of this Ocelot there is a tendency to dissociation of the rosette-spots.

Then in Fig. 19 (*b*) we have an Ocelot in which the same character is intensified and further modified, the enclosed spaces being of a different colour from either the rings or the general ground-colour.

In the Natural History Museum there are some Ocelots which show a further coalescence of the rosettes into more perfect longi-

FIG. 20.—Diagrammatic sketch showing transformation of Ocelot rosettes into longitudinal bands :—
(*a*) Rosettes arranged in longitudinal rows.
(*b*) Their upper and lower segments fusing.
(*c*) The rows of rosettes completely fused into bands of a brownish colour, margined with black.
(*d*) A row of rosettes from the flank of a Leopard skin ; these might readily fuse into *twin* stripes, as seen in Fig. 24.

tudinal bands, in which the traces of the Leopard rosettes are almost wholly obliterated ; and it would not be easy to conceive how they originated, without knowledge of other varieties of Ocelots which indicate the steps leading to the longitudinally banded Ocelot.

In Fig. 20 I have endeavoured to give a rough sketch of the passage from the Jaguar rosettes to the longitudinal parti-coloured bands of certain Ocelots.

C

The shading of *a*, *b*, *c* is intended to show the brownish colour of the enclosed space, so different from the general ground-colour.

In the same figure (*d*) I have given a row of rosettes taken from the flank of a Leopard skin, disposed diagonally. Like the Ocelot rosettes, they might readily coalesce into stripes. Indeed, there are many Tiger skins (Fig. 24) which have their stripes in pairs; and many brindled Dogs have similar markings. These may have resulted from such a disposition of rosettes as that here shown. I have seen several Leopard skins with their flank rosettes disposed in slanting rows (Fig. 15, *a*).

To understand the transformation of the Ocelot rosettes, we should bear in mind that the Jaguar rosettes are made up typically of a polygonal ring of spots enclosing a space which is of a darker colour than the *inter*-rosette spaces, and that the enclosed space contains some minute black *specks*. We find all these elements in the Ocelot markings, only they are differently arranged.[1]

Reference to the Ocelot figure on p. 417 of the *Royal Nat. Hist.* will make the transformation of the Jaguar rosettes into Ocelot bands quite clear. In this figure the lower row of flank marks is made up of *distinct* rosettes composed of distinct spots, like those of some variations of the Jaguar. The next row above it is largely made up of *fused* rosettes, and the row above that again is one long band of *fused* rosettes, the rings becoming the black border of the band, and the central spots becoming a row of small spots in the middle of the band. The rest of the body is covered with patches of fused rosettes. The Ocelot is essentially a South American species, and like its close relative, the Jaguar, is said to be an expert climber.

From the markings of the ocelot the transition is easy to those

See Stuffed Animals, Natural History Museum—case containing Ocelots.

of the Marbled Cat (*F. marmorata*), and those of the Clouded Leopard (*F. nebulosa*), both of which can be seen on pp. 409 and 407 of the *Roy. Nat. Hist.*

Besides the two variations of Ocelot which I have given, there are others which have longitudinal stripes that do not enclose any spaces, and are not unlike those on the shoulder of the light-coloured Ocelot in the foregoing figure (19).

There is a large number of spotted animals, and in the Appendix I have given as many as I could. The point the reader should note is that the rosettes of these spotted animals become dark *rings* on the tail, *alternating* with *rings* of the general ground-colour.

Having studied these rosetted Cats, we are now in a position to understand their relation to *striped* Cats.

There are numbers of small Cats and Genets with their spots arranged in longitudinal order, and others, as may be seen in Fig. 27, with them arranged in transverse order.

This happens frequently, not only in the Cat tribe, but in other animals also, as may be seen in Appendix A, Nos. 16, 17, 18, and others.

I have said that it will not be difficult to show that the striped animal is only a modification of the spotted one.

I have already shown that in the Ocelot spots run into longitudinal stripes and bands. Now I shall endeavour to show that the stripes of the Tiger are no other than the fusion of transverse strings of spots, so transformed as to have lost all semblance of their spot origin.

The striped animals *par excellence* are the Zebras and Tigers,[1] with the minor Tiger Cats. With the Zebras I shall deal in another place.

[1] In the *Animal Kingdom of Baron Cuvier*, Mr. Edw. Griffith, in vol. ii. p. 444, gives a pure white Tiger, with only a shading of stripes.

36   STUDIES IN THE EVOLUTION OF ANIMALS

Fig. 21 is the picture of a living Tiger, and Figs. 22 and 23 are the pictures of two very differently marked Tiger skins. It will be noted that on all three there are some spots which have not coalesced into stripes.

A glance at the spots of the living Cheetah and of the Cheetah skin (Figs. 10 and 11) will show that in many places they are arranged in transverse rows, viz., on the flanks and on the fore-legs; the Leopard spots on the fore-legs are also arranged in rows. A

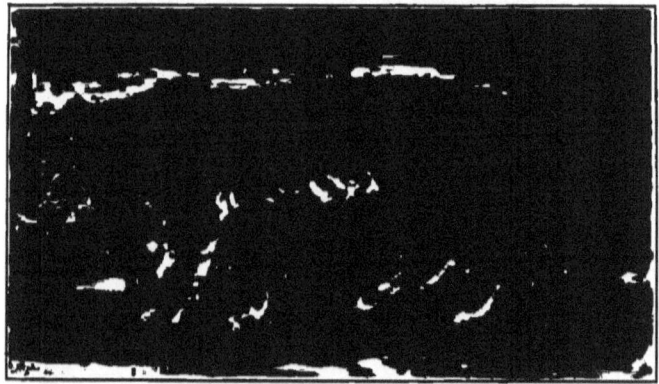

FIG. 21.—Picture of a living Tiger, from a photograph by Ottomar Anschütz.

very little change will make the spots closer, and a further change will first amalgamate them into beady strings[1] and then turn them into bands, as we have seen occur in the Ocelots; only in the Tiger and certain Cats the bands are transverse and their margins are sharp, while in the Ocelot they are longitudinal and their margins are scolloped.

It is very strange that spots in certain animals, and in certain

[1] In a previous note I mentioned having seen Cheetah spots run into beaded strings.

SPOTTED AND STRIPED MAMMALS   37

parts of the same animal, very frequently show a tendency to arrange themselves in either longitudinal or transverse orders, and often coalesce into stripes. When irregularly disposed rosettes

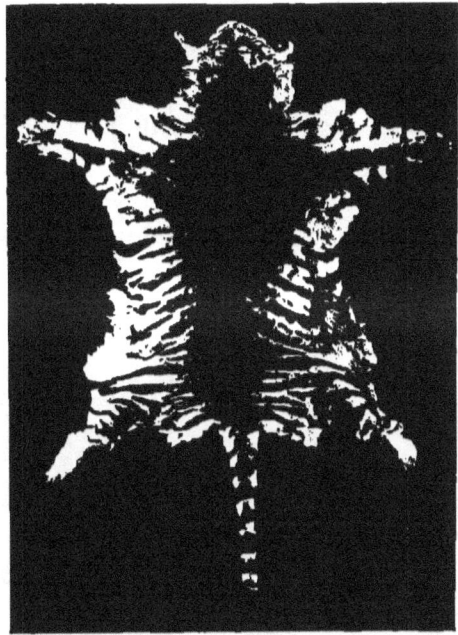

FIG. 22.—Tiger skin, from a photograph by Messrs. Dixon and Son. Skin lent by Messrs. Jeffs and Harris.

coalesce, they form the large patches of the Clouded Tiger (*F. diardi*) of Elliot's monograph.

What is still stranger is that the black rings of the rosettes

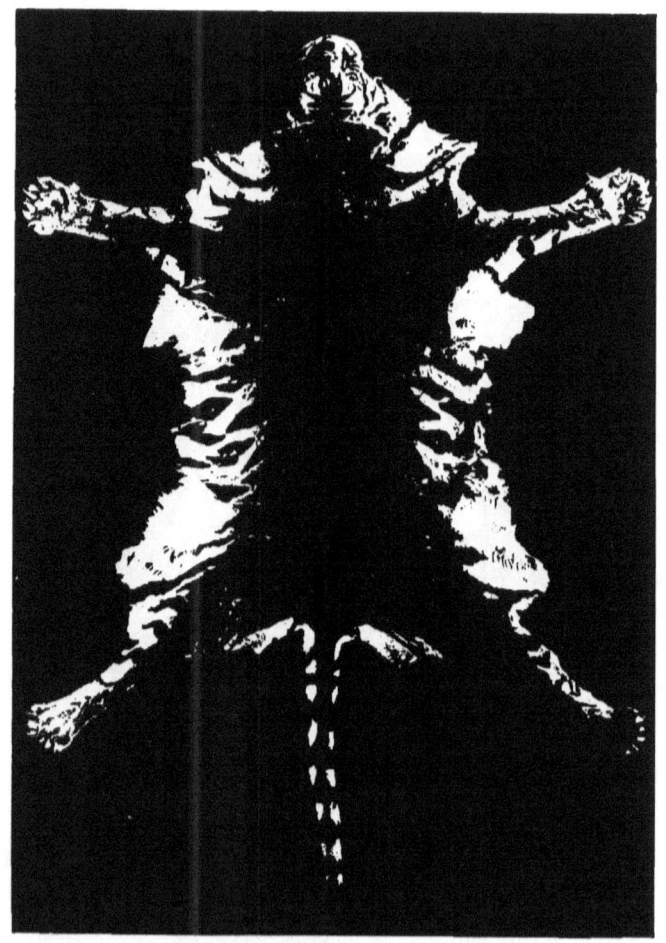

FIG. 23. Tiger skin, from a photograph by Messrs. Dixon and Son. Skin lent by Messrs. Jeffs and Harris.

coalesce with other black rings, and form a larger ring or band *outside* the amalgamation, and the brownish enclosed spaces

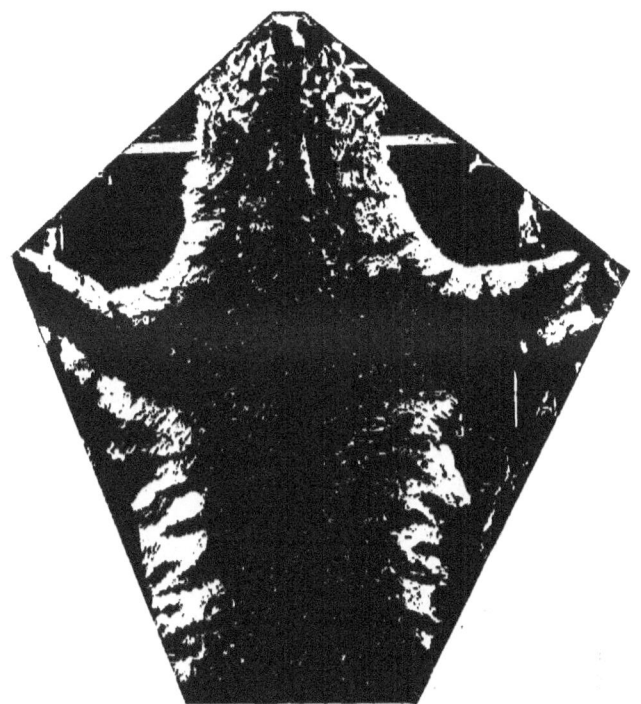

FIG. 24.—Tiger skin showing twin stripes. Photograph obtained from Messrs. Russ and Winckler, of Edinburgh.

coalesce and remain *inside* the patch or band, as in the Clouded Tiger and in the Ocelot.

40 STUDIES IN THE EVOLUTION OF ANIMALS

Fig. 25 shows various degrees of transformation of spots into stripes and blotches on the legs of certain carnivora. On the other hand, we should not forget to note that on the legs of striped

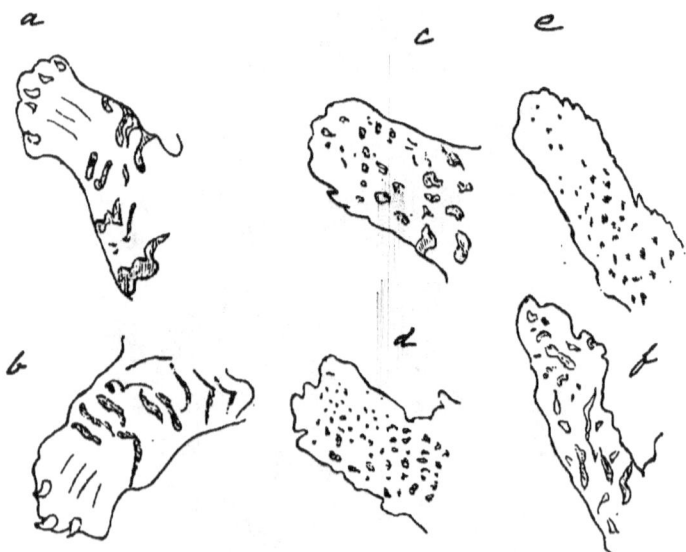

FIG. 25.—*a* and *b*, fore- and hind-feet of a tiger; *c* and *d*, fore-feet of two different leopards; *e* and *f*, fore-feet of two different lynxes. From *Coloration of Animals and Plants*, by Alfred Tylor.

animals sometimes all markings disappear. In the window of a furrier I have seen a young tiger with plentiful stripes on its hindquarters and almost none on its shoulders and fore-legs; and we know also that the Quagga (Fig. 54) is plentifully striped on its front parts and has no stripes at all on its haunches and hind-legs.

SPOTTED AND STRIPED MAMMALS   41

I have seen a Leopard skin with crowded rosettes, several of which were fused into one oblong ocellus, as shown in Fig. 15, *b*.

This fusion will afford some idea of how the parallel twin stripes of the Tiger skin of Fig. 24 may have originated. Fig. 26 gives a diagrammatic sketch of some spindle-shaped ocelli I saw on a Tiger skin.

A row of Leopard rosettes, as shown in Fig. 20, *d*, may easily be transformed into a pair of twin and parallel transverse stripes.

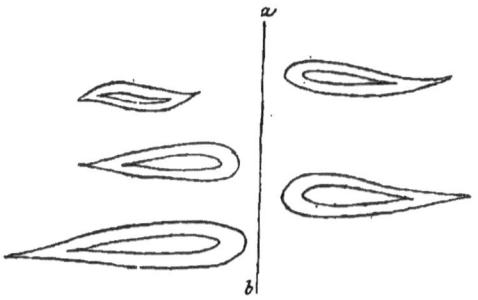

FIG. 26.—Spindle-shaped ocelli on a Tiger skin; *a—b* represents the spinal line.

We have already seen that similar twin bands do actually occur in the Ocelots from rosettes disposed longitudinally.

The reader should especially note that in the figure of the living Cheetah the spots on the tail have gradually coalesced, and formed continuous rings towards the tip. These ringed tails occur whether the animal be spotted or striped, and in the figure of the living Tiger the rings on the tail are *double*, indicating their origin from rosettes; and indeed in Fig. 24 almost all the stripes on the body are double. This may be considered conclusive evidence that the stripes of all Tigers, however modified they may be,

42 STUDIES IN THE EVOLUTION OF ANIMALS

originated from *rows of Leopard rosettes*. We may note further that in the Jaguar skin (Fig. 4) the spots on the chest have

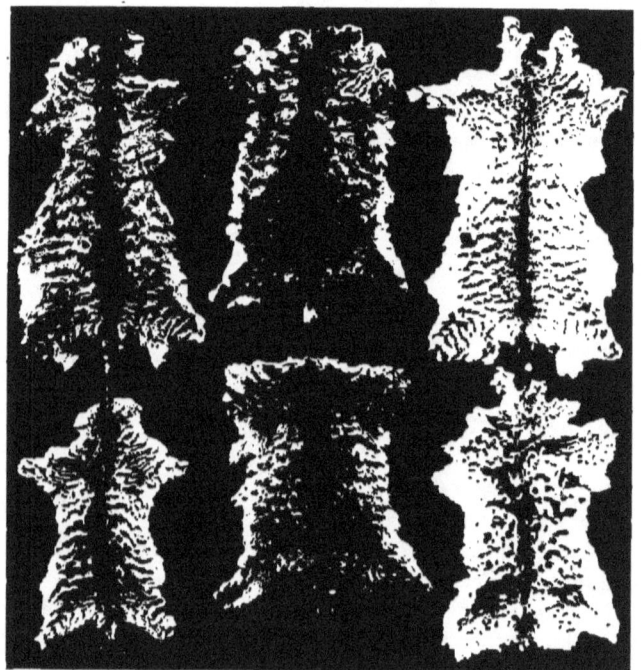

FIG. 27.— Markings of different small Cats, from a photograph by Messrs. Dixon and Co. Skins lent by Messrs. Back and Co.

coalesced into beady stripes, and we know that the separate ring-spotlets of the Jaguar rosettes in other Leopards do coalesce into a continuous polygonal ring.

SPOTTED AND STRIPED MAMMALS 43

Then in Fig. 27 we have a striking series of transitions in the smaller Cats from spots to stripes. Some specimens show only spots on the flanks; others, spots coalesced into wavy or beady stripes; and others show more finished stripes, like those of Fig. 28. In the British Museum enclosure there is a Domestic Cat which I often stop to look at. The posterior half of its body is spotted; the anterior half is striped transversely; the neck and head are striped longitudinally; the legs are striped transversely, and also spotted. Then along its spine it has a broad black band, and its tail is ringed in its terminal half. Here we have a sort of generalised marking, combining a little of each of the special features of distinct races of animals, the black band along the spine in some animals being possibly the *only* vestige of ancestral spotting; while the ringed tail in the Racoon is the *only* vestige left to tell the tale of its ancestral markings.

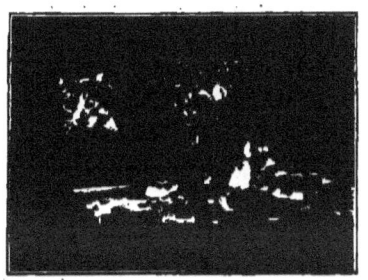

FIG. 28.—Striped Cat, from a photograph by Messrs. Dixon and Son. Near the root of the tail it has a few *rosettes*.

In the International Fur Stores I saw the skin of a Tiger which had a large ocellus towards the ventral region. This same skin had a modification of stripes on the lumbar region, as if the pigment were undecided whether it would run into separate stripes or form an ocellus. Then in another Tiger skin I saw on one side a curious rosette, and on the other a pair of parallel stripes. Both these abnormalities in the Tiger markings are given in Fig. 29 (No. 1). They are not only curious, but very suggestive.

If one had the opportunity of examining hundreds of skins of Leopards, Cheetahs, Tigers, and Cats, I have no doubt whatever that a perfect series might be picked out which would easily prove the transition from spots to stripes. The examples which I have given, I think, are sufficient to convince any one of the relationship of stripes to spots and rosettes.

A word about the tadpole-shaped ocelli on both sides of the

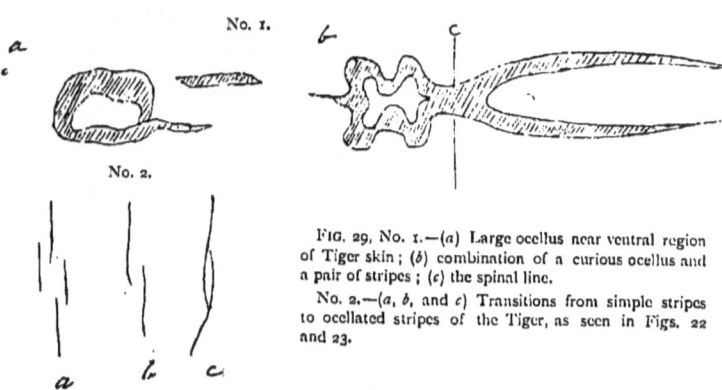

FIG. 29, No. 1.—(a) Large ocellus near ventral region of Tiger skin; (b) combination of a curious ocellus and a pair of stripes; (c) the spinal line.

No. 2.—(a, b, and c) Transitions from simple stripes to ocellated stripes of the Tiger, as seen in Figs. 22 and 23.

Tiger skin of Fig. 23. I do not think that these are enlarged and spindle-shaped Leopard rosettes, although in Fig. 59, No. 5, I have given one which has turned into a *beaked* ocellus. I think that the Tiger marks in question have a different origin. Tiger stripes are sometimes parallel throughout their whole length, as seen in Fig. 24; but at other times the stripes are shifted so as to make one commence about the middle of the other, as in the lumbar region of Fig. 21. By approximation and partial fusion, two stripes thus shifted would make a spindle-shaped figure, with an ocellus in the middle; if the ocellus were closer to one end, the figure would become not unlike the outline of a tadpole, like those on the Tiger skin of Fig. 23.

No. 2, Fig. 29, is a diagram intended to show the transformation from simple stripes to spindle-shaped ocelli.

I think I have said enough (and perhaps more than enough) regarding the striping of Tigers to show that it is simply an extreme modification of the Jaguar and Leopard rosettes. But if the reader should have any doubts about the descent of stripes from spots, a glance at the small Cat skins of Fig. 27 ought to convince him that the view I have taken of the genesis of stripes, in the Cat tribe at least, is in all probability the *right one*. We see separate spots passing into beady stripes and finally into Tiger stripes on the hind-legs. On the shoulders of the small Cats the striping is so fine that it is rather a brindling.

Then in the Natural History Museum, among the Cat family, there are numerous specimens which show simple spots, mixtures of spots and stripes, and simple stripes either transverse or longitudinal, and also transitions from the one kind of marking to the other.

I might have dispensed with such a multiplicity of facts in support of what I said; but to the general reader, who may not be in the habit of seeing at a glance the obviousness of a conclusion, they may be useful in bringing home to him the truth that *stripes are evolved from rosettes*.

If now we turn to other animals, such as the Deer and the Antelope, we shall find that spots and stripes are interchangeable and intermixable.

Fig. 30 shows the spotted young one of a Deer in the Zoological Gardens; the adult showing no spots whatever.[1]

---

[1] In the Edinburgh Museum of Science and Art there is a good specimen of a young spotted Wapiti (*Cervus Canadensis*).

46 STUDIES IN THE EVOLUTION OF ANIMALS

Then if the reader will turn to the *Review of Reviews* of March 15, 1893, where a character-sketch of the great African hunter, Mr. F. C. Selous, is given, he will find on p. 258 a Kudu Bull beautifully striped, with no spots, and on p. 260 an allied Antelope, the Bushbuck of the river Chobe, which is striped and also spotted. Of the latter a fine illustration is given in

FIG. 30.—Shows the young of a Deer covered with spots, while the adult has none; taken from a photograph, Zoological Society's Gardens.

the *Roy. Nat. Hist.*, vol. ii. p. 277. See also Appendix A, Nos. 20, 21, 22, 23, and 25, of this book.

The changes from spotting to nothing in the same individual, and from spotting to striping, or a mixture of both, in the same genus of Antelopes, is very remarkable. For some reason the spotting of the adult of *Axis maculatus*, shown in Fig. 14, is not obliterated, although in the Deer of Fig. 30 no sign of spotting remains in the adult.

It will, however, be preferable to let Mr. F. C. Selous[1] speak upon the question of spotting and striping in the Antelopes of Africa. It would appear that not only age, but the climate and food of a locality, may have a good deal to do with the changes in the markings of the skin.

Of the Bushbuck (*Tragelaphus sylvaticus*, Sparrm.) he says: 'In the Cape Colony, the adult male is deep dark brownish-black, with two or three small white spots on the haunch, and one or two on the shoulder; the adult female is light reddish-brown, with white spots on the haunches, and sometimes a few between the shoulders and flank; the young males are reddish-brown, and more or less spotted.

'On the Limpopo, however, the adult males are brownish-grey, often *without a sign of any spots*; and the young females are more spotted than old ones. The adult females are of a dark red, with a few white spots; the young males are a good deal spotted, with a few transverse stripes.

'On the tributaries of the Zambesi, east of the Victoria Falls, the male Bushbucks are of the same colour as the young males found on the Limpopo, being dark red thickly spotted on the haunches, shoulders, and sides, with *small* white spots, and with three or four faint white stripes down each side. The adult female is pale yellowish red, beautifully spotted, and with a few white stripes.'

Then of the Bushbucks on the Chobe Mr. Selous says: 'The adult males are of a very dark red colour, most beautifully spotted with *large* white spots, as many as fifty on each side in some individuals, and in some cases as many as eight well-defined stripes. In addition, they have a mane of white hair three inches long, from shoulder to tail, which can be erected. Young

[1] *A Hunter's Wanderings in Africa*, p. 209.

males are of a pale red-yellow, with spots and stripes much more faintly marked than in the adult animal. The adult female is of a rich dark red, beautifully spotted with white, and with three or four faint white stripes on each side, and a dark spinal line. The young female is of a lighter red, and not so much spotted. In Cape Colony and on the Limpopo, young Bushbucks are more spotted than adults; they gradually lose their markings as they become older; while on the Chobe and on the tributaries of the Zambesi this order of things is *reversed*. Adult animals are far more beautifully marked than young ones.'

Then of *Tragelaphus Spekii* (Sclater) he says: 'A fœtus had the skin striped and spotted yellowish-white, as in the adult Bushbuck of the Chobe. Another recently born had a lighter colour, and fainter spots and stripes; the adult is of a uniform greyish-brown *without* either spots or stripes.'

From all this it would appear that the spotting and striping of the Bushbucks does not depend on either age or sex, and that the individuals of this genus differ much in marking and colouring; but, although differing so much as to become almost distinct species, physiologically they remained *one* species; and, living together in the African bush, they must have crossed and have become mixed up, so that the life-history of *one* individual seems somehow to give successive photographs, as it were, of the life-history of the *race*.

Anything more bewildering than the facts placed before the reader by this hunter-naturalist of Africa cannot well be imagined.

It is impossible to read Mr. Selous' book and not feel convinced that these Antelopes, when living for months in *waterless* tracts, are at the mercy of their surroundings, not only for their life, but also for the physiology of their skins. For at p. 207 he says that in some parts of the country, for several months in the year,

SPOTTED AND STRIPED MAMMALS

there is absolutely no water, and Elands, in common with Gemsbuck and Giraffe, live and thrive there, and appear to do better than in well-watered parts of the country. He thinks that in the dry seasons these desert animals get the necessary amount of water by eating a watery melon, which is plentiful, and a white watery bulb, looking much like a turnip.

Among Antelopes, we have the head of a *Kudu* with *three white spots* on its cheek.[1] Then in Appendix A, No. 27, is shown *one* solitary broad stripe on the hind-quarters of the Waterbuck. These, like the ring-tails of many animals, I take to be simply *vestiges* of more extensive ancestral spotting and striping. Indeed, climate, food, and age may have a great deal to do with the retention or disappearance of spots and stripes. We see that in the Deer of Fig. 30 the spots *wholly* disappear with age, while in these two Antelopes all spots and stripes disappear, excepting, maybe, three cheek spots in the one and one haunch stripe in the other.

Mr. Selous says: 'In the Mashura country every Eland cow is plainly striped. One had nine broad white[2] stripes on each side. Elands that are much striped have a whitish mark across the nose, like the *Kudu*. Old bulls have no stripes. Great variations occur in this respect.'

Again turning to Dogs, we find that spotting and striping are found among them also.

Fig. 31 shows a distinctly spotted Dog.[3] Whether the Dog got

---

[1] Shown in Mr. Selous' book, in vol. ii. figs. 1 and 2.

[2] It should be noted that in certain mammals, such as the Tiger and Zebra, the stripes are *black*: while in these Antelopes the stripes are *white*. Both spots and stripes are liable to change from black to white, or *vice versâ*; that is, white becomes *melanoid*, and black becomes *albinoid*.

[3] In the Natural History Museum there is a Phalanger—a marsupial—spotted much in a similar manner. In the *Encyclop. Brit.* the picture of a flying Phalanger is given with transverse stripes on its back. In Somerset I saw a sucking Pig which was spotted almost like some Dalmatian Dogs, but the spots were larger.

this spotting from some mammal allied to the Cheetah, or not, I do not know. Canine and feline animals are rather closely allied. Moreover, both the Cheetah and the Dog want the retractile claws of the Cats proper.[1]

FIG. 31.—Dalmatian Dog, from a photograph by C. R., 54.

Fig. 32 shows the Dog-spots fused into blotches not unlike the consolidated rosettes on the abdomen of Leopards. In the Science and Art Museum of Edinburgh there is a large Boarhound with the Dalmatian spotting agglomerated into even larger blotches.

[1] Among Hyænas there are species which are distinctly spotted, and others distinctly striped; although in Hyænas the claws are well developed, they are non-retractile.

Probably this is only a preliminary step to piebalding, which is so common in Dogs. It is conceivable that if this attraction of black spots for each other increased, we should soon get the piebald colouring of the Fox-terrier and other breeds.

At Worthing I saw a curious hybrid between a Bloodhound and a Dalmatian Dog—at least, its owner so stated. Its legs were white with well-defined tan spots; its back and flanks were white and thickly spotted with black; and those on the flanks were grouped in threes and fours, not unlike the consolidated rosettes on the abdomen of the Jaguar and Leopard; while its head was white, spotted with tan. It showed a curious persistence of spotting; but where the tan colour of the Bloodhound came in, the spotting of this Dalmatian hybrid became tan-coloured!

FIG. 32.—Blotched Danish Boarhound, from a photograph by Messrs. Dixon and Son.

Then in the streets of London I saw another Dog of the Dalmatian breed. A large number of the dorsal spots amalgamated into a large black patch; half its face was spotted as usual, and the other half wholly black.

On another occasion I met two Dogs of this same breed, probably brothers. One was almost wholly black, with some remaining white inter-spots, and half its tail was spotted with black;

52 STUDIES IN THE EVOLUTION OF ANIMALS

while its companion had the black spots quite close to each other, though much more distinct.

Then in Hyde Park one day I saw a curious Toy-terrier. It was of the black-and-tan, short-haired breed; but the parts that are usually wholly black, in this case were *grey*, blotched and striped

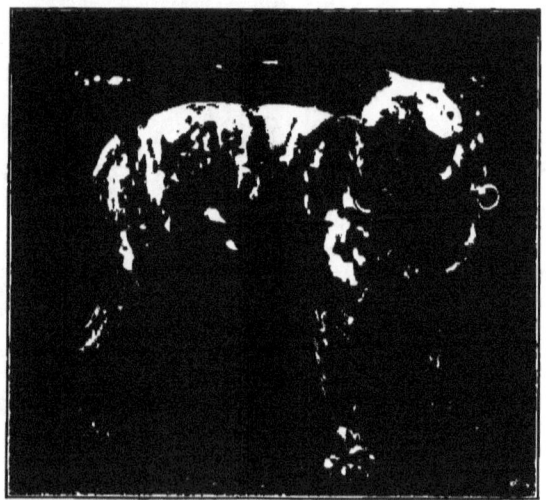

FIG. 33.—Brindled Dog, from a photograph by C. R., 847.

with black. It was not unlike the black-backed Jackal of the Zoological Gardens.

Fig. 33 is that of a Brindled Dog. Here the stripes are unlike those of the Tiger—that is, not so decided, but more like the brindling on the shoulders of the small Cats in Fig. 27. The stripes of the Dog are much finer and dissociated; but in the streets I have seen brindled Bull-terriers which had closer and broader stripes, with some which were broken up into fine lines; and in some cases many of the stripes were distinctly in *pairs*, suggesting some origin similar to those on the Tiger skin of Fig. 24, and presumably therefore the twin stripes in the Dog must have been caused by a similar modification of rosettes.

Mr. Rawdon B. Lee, in his book on *Modern (Sporting) Dogs*, gives a good figure of a Danish Boarhound with conspicuous stripes.

In brindled Dogs we have to note that the stripes on the limbs follow the direction of the limbs, and are not transverse to them, as in Cats, Zebras, and striped Hyenas. In some Tigers I have seen a tendency to a similar disposition of the limb stripes. This, however, need not embarrass us, as we have already seen that, in the Cats, spots may group themselves into stripes either longitudinally, or transversely, or diagonally. Brindled Dogs are to be found in various races—in Greyhounds, Boarhounds, Bulldogs, etc.

In the *Viverridæ* there is presented a similar study. We see large spots breaking up into numerous small ones, or perhaps the reverse[1]—that is, small ones agglomerating into large ones; also spots stringing themselves into stripes, longitudinally on the body, and transversely on the legs.

In several species of the Mungoose a complete intermixture of pigments seems to have occurred, so as to produce a sort of grizzly-grey coat; while in the Zebra-Mungoose and in the banded Mungoose, both of East Africa, the spots have arranged themselves in stripes and bands transversely (see App. A, No. 15). In *Galidea*, however, we see a wholly brown surface, with a few black rings on the tail, as a vestige of ancestral striping or spotting.

What we have to note very particularly is that the Indian Civet

[1] See Appendix A, Nos. 9 to 12.

(*Viverra zibetha*) in the Science and Art Museum of Edinburgh has *Leopard rosettes* on its haunches, simple spots on its shoulders, and marblings on its flanks! and that the Spanish Lynx (*Felis pardina*, Oken) has *Leopard rosettes* on its haunches, and simple spots over the rest of its body.

After having passed in review so many spotted and striped mammals, we may perhaps be in a position to divide their markings into—

(*a*) Spots and groups of spots forming rosettes, or solid blotches.

(*b*) Stripes or bands of various breadth, either transverse, diagonal, or longitudinal.

(*c*) Marbled or clouded markings, like those of some Cats.

(*d*) Piebald markings, like those of Dogs, Cattle, Horses, etc. And finally we have

(*e*) Self-coloured animals, which present a total obliteration of spotting and striping. These may be subdivided into pure selfs, without a speck of any other colour; selfs with vestiges of spotting; selfs with ringed tails; and selfs with points of other colours, such as we see in Horses, Dogs, etc. The dun-coloured Cat, with black points, is a very interesting variation. It is seen in exhibitions.

From the study of all the foregoing, I have come to the conclusion that each ancestral rosette, originally composed of a ring of isolated spots, has in time undergone the following marked modifications, some of which are found in the same animal, and others in distinct individuals :—

(*a*) The isolated spots have fused into continuous rings, or segments of rings, as in many Leopards.

(*b*) The ring has contracted into a large spot, with obliteration of the enclosed space, as in the Serval and others.

SPOTTED AND STRIPED MAMMALS

(c) The rosettes, sometimes many of them, have fused either into large bands or patches, as in the Ocelot and marbled Cat.

(d) The consolidated spots, after arranging themselves into rows, either transversely, diagonally, or longitudinally, have fused themselves further into stripes, as in the Tiger, the 'Tabby Cat,' the Pampas Cat, and certain Civets.

(e) The rings on the tail have followed the same course of modification; the consolidated spots have fused into rings, or the rosettes have fused into twin rings, as in the Margay; these, in some descendants, have then amalgamated into broad bands.

(f) Finally, the rosetting, spotting, or striping has been entirely obliterated from the adults of certain species, such as the Lion, the Puma, the Caracal, jet-black and albino Cats, and others, while in the young of some the spotting remains distinct.[1] The rusty-spotted Cat 'is quite peculiar among spotted Cats in having the tail without either spots or rings;' while in some mammals, as in the Racoon and the cunning Bassaris, the only vestiges of ancestral spotting or striping are the 'rings on the tail.'

[1] Young Lion cubs are usually spotted; but Mr. Edward Griffith, in *The Animal Kingdom of Baron Cuvier*, vol. ii. p. 447, gives two Lion-Tiger cubs, three months old, striped like Tigers.

DAPPLED AND STRIPED HORSES

AND SOME OTHER MAMMALS

'The first master strikes out a luminous idea, and writes a great book which promises speedy results; but after his own generation has been dazzled by it, comes the criticism of the next: exceptions, and violations of his laws, are discovered; the large views which he stated with convincing clearness become misty and obscure; and men set themselves to rediscover, in some new way, generally with poor and shabby minuteness, and with many modifications, what was once an accepted theory.'

The Present Position of Egyptology, by Professor MAHAFFY,
Nineteenth Century, Aug. 1894, p. 269.

PART II

DAPPLED AND STRIPED HORSES

AND SOME OTHER MAMMALS

FOR my purpose, under the term Horse I include all animals that come under the denomination of the genus *Equus*.

In the Leopards and Tigers it was easy to show the derivation of stripes from spots or rosettes. There is, however, a domestic animal which is differently marked from Leopards. I mean the dappled grey Horse shown in Fig. 34. He has a congener—the Zebra—in which the stripes are quite phenomenal. In these animals it is not so easy to trace the striping from spotting, although, I think, it can be done.

There are three distinct varieties of *fully* dappled Horses like that shown in Fig. 34, viz., the white Horse reticulated with grey, the dun Horse reticulated with black, and the brown Horse also reticulated with black.[1] The markings of the dun or sponge-coloured Horse are very striking. They are all called *dappled* Horses in the trade—grey, dun, brown. I have called the darker pigment *reticulation*, because it appears as if a net were thrown over the fully dappled Horse, leaving the meshes filled with either white, dun, or brown.

The dappling or spotting of the domestic Horse is so persistent,

[1] Not improbably the dun and the brown dappled Horses are *melanoid* variations of the grey dappled Horse.

FIG. 34.—Dappled grey omnibus Horse, from a photograph by Mr. Gabrielli (taken by kind permission of Mr. D. Duff, manager, London Road Car Co.).

that we see it, more or less, going through all the changing colours of Horses; and it would be almost as hopeless to give distinct names to all the variations of colour in Horses, as it would be to name all the shades of colour in domestic Pigeons. The very fact that the dappling is so persistently inherited, either wholly or vestigially, would indicate that it comes from the very foundations of Horse evolution.

I have not been able to discover that any existing species of the genus *Equus*, in the *wild* state, is dappled.

The earliest record of a dappled Horse, in a state of domestication, that I can find is taken from a Spanish MS. of the eleventh century. The quaintness of its marking is shown in Fig. 35, and the author thinks it was of the Arabian breed. The markings are most conspicuous in the grey dappled Horse, because the colours are in black and white; but the dappling is traceable in the bay,

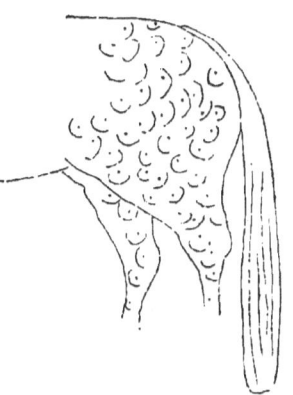

FIG. 35.—Spanish Horse, from *Horses of Antiquity*, by Ph. Ch. Berjean (p. 23, MS. xi. Cent.).

the chestnut, the brown, the black, the roan, the cream, the dun, etc. etc. The pure white, the pure black, and other pure self-coloured Horses may be free from traces of dappling; but the vast majority of Horses are either fully dappled, or have *traces* of dappling, and these are most persistent on the hind-quarters, round the root of the tail. The Horses of the 2nd Life Guards are either black or nearly so. I noticed that those which took

62   STUDIES IN THE EVOLUTION OF ANIMALS

part in the escort at the opening of the Imperial Institute were almost all dappled on the hind-quarters, round the root of the tail.

In dun [1] or sponge-coloured Horses, the dappling in this particular region is often amalgamated into a sooty-black patch on

FIG. 36.—Dappled grey Cart-horse, from a photograph by Mr. Stanborough, of Bexhill-on-the-Sea.

each side of the tail-root; a similar sootiness results from the fusion of dappling along the upper ridge of the neck.

Some grey Horses are very distinctly and strongly dappled, such as those given in Figs. 34 and 36. I think it very beautiful

[1] Mr. Darwin defines *dun* as ranging from 'between brown and black to a close approach to cream-colour' (*Origin of Species*, 1888, p. 200).

DAPPLED AND STRIPED HORSES 63

to see a pair of finely shaped grey carriage-Horses prettily spotted. Should any of them die, their skins would make pretty rugs. Unfortunately, for some reason, spotted Horses are not fashionable, and do not appear to be kept for breeding purposes.

In some instances the dark reticulations are all over the body, but more conspicuous on the flanks, the dappling on other parts being often modified. The white dapples of the grey Horse can be seen to vary from irregular patches to small spots. In some regions, such as on the fore- and hind-legs of Fig. 36, the smaller dapples seem to have amalgamated into large fern-like patches. In other cases, such as those on the fore-leg of Fig. 34, the dapples have degenerated into star-shaped marks, which often dwindle into minute specks like those on the flank of Fig. 39, all the rest of the ground being of a dark grey, and in some cases almost black. Indeed, we might say that the two extremes of the grey dappled Horse series were—(*a*) a *white* Horse with traces of grey reticulations, and (*b*) a *blackish* horse with traces of white spots. In other words, the grey reticulations that isolate the white dapples of the grey Horse can be obliterated, either wholly or partially, and the Horse made either wholly *white*, or white with *vestiges* of dappling. On the contrary, the white dappling may be obliterated, and the Horse made either *wholly* dark grey or dark grey with *vestiges* of white spotting. The star-marks on the fore-legs of Fig. 34 should be compared with those of Fig. 36.

The invasion of the white dapplings by the grey reticulations is partially seen in Fig. 37.

Few, I venture to say, have any notion how much may be learnt from Horses of all sorts which are to be seen by thousands in the streets of the Metropolis. They are ready-made experiments for scientists to take up and theorise about. Unless pure white,

64   STUDIES IN THE EVOLUTION OF ANIMALS

chestnut, or flea-bitten, there is scarcely a Horse which does not bear vestiges, more or less pronounced, of dappling.

In Fig. 38 is given a spotted Horse, which shows the dappling on the flank disposed in *slanting rows*; and Fig. 34 shows that on the neck the dark colour tends to form *bands*.

FIG. 37.—Grey dappled Pony, from a photograph by C. R., 922.

Note that on the groin of both Horses of Fig. 38 the dapples are being broken up into *minute specks*.

When this occurs all over the surface, it probably gives rise to what is called a 'flea-bitten' Horse, with either black or brown specks.

DAPPLED AND STRIPED HORSES

I would ask the reader to note particularly the *rows* of slanting dapples on the flank of the upper Horse of Fig. 38. I have already referred to similar slanting rows of rosettes in certain Leopard skins (Fig. 15, *a*), and shall have to refer to them again.

No one, I think, will doubt that these spots in this Horse are *vestiges* of the larger dapples, like those of Fig. 34. If a full-blown dapple or rosette can be reduced to a mere point, as we see it in certain Horses and other animals, it stands to reason that it can be much modified otherwise.

We do not know what atomic conditions of the nerve-centres are requisite to produce this minute specking, but it is evident that a more complete intermingling of the pigmented hairs with the white ones would give rise to a roan or a strawberry roan. On the other hand, when the pigments agglomerate in separate large patches, we get the same piebald conditions seen in Cattle, Dogs, Pigs, etc.

I am not here going to enter into the intricate question of how the dappling of the young Horse commences—whether by minute spotting becoming larger, or by large patching dwindling eventually into minute spotting. It is a difficult question to unravel, and there does not seem to be any accurate information on this point. As to how the dappled Horse originally came into being at all, there would not seem at first any means of finding out. Fossils in no way record the coloration and markings of the skins of extinct animals. Nevertheless, I hope to throw some light on this point of evolution later on. The existing wild congeners of the Horse are either striped, like the Zebra, or self-coloured, like the *Kiang* or wild Ass. The coloration of the latter is a sort of tan or fawn colour, while in the domestic Ass mouse-grey is a common colour.

The person who originally invented the names for the colours of Horses must have been colour-blind, for how could he have called

FIG. 38.—Two Cart-horses, from photographs by Mr. P. D. Coghill of the Royal Veterinary College, taken by kind permission of Mr. J. Poynter, Horse Department, Great Northern Railway Company.

'chestnut' the colour of a Horse, which, if it were a Cow, would be 'red.' And to call a Horse *tan-coloured* would be simple heresy. Nevertheless, what is called a golden-bay is nothing but a rich tan-colour, which in other individuals shades off into the dun, the sponge-colour, the cream, and the white; while in another direction this tan shades off into the chestnut, the bay, the brown, and the black.

All the experts whom I have consulted agree in saying that the Horse, when recently foaled, is never dappled, but is of a uniform dark colour, excepting albinos.

I asked a farmer who is a great hunter, and who has also bred Horses, and has had ample opportunities of seeing young foals, whether he had ever seen a recently born foal which was dappled. He replied—not one. I am informed that the dappling, when it does come, begins to appear when the foal is a few years old. This is rather curious, for in the case of some Deer, as shown in Fig. 30, the young one is plentifully spotted, while the adult has no sign of spotting.[1]

No one seems to have made any accurate observations on the dappling of the Horse—when it commences, how it commences, and how it proceeds, although there are many records of Zebra-striping in the Horse. Among the thousands of Horses in the streets of London, one sees all possible variations of dappling—from a few spots to the whole body covered with maculations. Does the same individual go through all these phases of dappling, or are certain variations permanent? Does dappling commence gradually and go on to its maximum extent, and then gradually disappear, or how? A veterinary surgeon told me that the dappling varies with every change of coat of the Horse; and all seem to agree that as the Horse grows older the white colour increases and the dark colour

[1] In the Natural History Museum there is a very young Jaguar which is wholly *brown* without any spotting.

68   STUDIES IN THE EVOLUTION OF ANIMALS

diminishes. The chafing of the collar seems to have the same effect; but in Fig. 39 and many others the reverse seems to have occurred. Does sex make any difference in the dappling? A

FIG. 39.—Omnibus Horse with the spots becoming obliterated, leaving a dark-grey surface; from a photograph by Gabrielli, taken by permission of Mr. Duff, manager of the London Road Car Company.

great deal is known about the powers of running of the Horse, about his powers of draught, about his anatomy, his physiology, his pedigree, and perhaps about his descent and relation to extinct and living animals; but very little seems to be known about the origin and course of his dappling.

DAPPLED AND STRIPED HORSES 69

What is wanted is that some amateur with leisure and means should undertake to photograph the dappling of the same Horse as soon as this feature commences, and to photograph the *same Horse* on both sides *every year*, after the Horse's change of coat. So many leisured persons possess cameras, that this bit of work ought to be an easy amusement.

If many were to undertake this work, so much the better. And if the Horse were *closely* clipped after being photographed, and then re-photographed, we might be able to discover how far the pigmentation of the skin coincides with the dappling of the hair.[1] We would then have some accurate data to build theories upon ; and by comparing the dappling year by year, we might have the revelation of some interesting facts. All these seem trivial things, but from an evolutionary point of view they may be important. However trivial a fact may appear at one time, it may one day turn out of value.

This any one can see for himself, viz., that the colour of the clipped Horse is much lighter than it is before clipping. Part of the darkness of the old coat may perhaps be put down to dirt, and part to the action of light and weathering. In some cases, after clipping, the surface shows traces of dappling which it did not show before ; and in roan Horses the clipped surface is often darker, or lighter.

We seem to have more definite information about crustaceans, fished out of the ocean from a depth of 3000 fathoms, than we have of the changes in the colouring and marking of the Horse, an animal which has been in daily use, in various ways, for thousands of years !

---

[1] I saw a pink-white Horse in an omnibus. It had very little hair—indeed, it was almost hairless ; and on its shoulder it had dark pigment-marks in the *skin*, like those of the hair-marks on a grey dappled Horse. Moreover, it had dark circular spots in many parts of the skin, like those of the Dalmatian Dog, the rest of the skin being of a blush-rose colour.

Considering the millions of Horses that are bred everywhere, both in a domesticated and also in a semi-wild state, it is astonishing how little reliable information there is on this particular subject. As I said, from an evolutionary point of view this study would be important, as the history of dappling in any individual ought to tell some tale regarding descent.

In absence of any recorded accurate information, I have had to explore the Horses in the streets of London and elsewhere, in the immense stables of carrying companies, in Horse-shows, etc. Wherever I could see any deviation which I considered of some importance, I made a note of it on the spot.

Have you ever observed a 'dappled sky'? The maculations of some dappled Horses, such as those of Fig. 37, are not unlike the patches of cumulus clouds with jagged edges, as they are often seen close to each other in a 'dappled' sky, the intervening blue sky corresponding with the network of dark grey between the Horse-dapplings. Just as clouds are unstable, break up, and run into each other, forming a uniform dense haze, so do the jag-margined Horse-dapplings seem to change with age and from other causes. The cloud-patches of the Horse often break up and deliquesce—to continue the simile—as seen on the hind-quarters of Figs. 34 and 39.

This I found, and have confirmed it hundreds of times in omnibus Horses. After severe exertion in drawing a loaded omnibus, the superficial veins of the flanks, shoulders, and legs start out. On the flanks they form reticulations which, in dappled or partially dappled Horses, *coincide* with the dark reticulations. In self-coloured Horses, such as pure whites, pure bays, etc., only the *venous* reticulations are seen. But whenever there is a fine pigmentary reticulation of a different colour from the general colour, the two reticulations *coincide*—the venous and the pigmentary. In freshly clipped Horses all this can be seen very plainly.

DAPPLED AND STRIPED HORSES

In another place I have endeavoured to account for this coincidence.

In Fig. 40 (No. 1) I have given a diagram of what might be a portion of the flank veins of a Horse; and in No. 2 I have shown the same pigmentary reticulations *broadened* by invasion of the dark pigment, or, what comes to the same thing, by contraction of the white patches. The broad reticulations still correspond with the venous network, shown by dotted lines, which lies in the

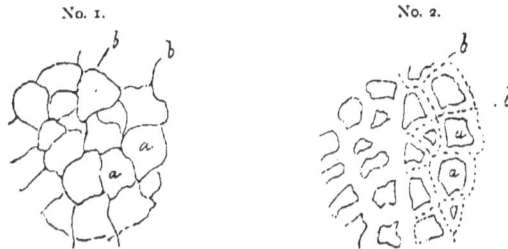

FIG. 40.—Diagrams of reticulations of flanks of Horses.
No. 1. (*a*) white patches; (*b*) dark reticulations, coinciding with veins.
No. 2. (*a*) contracted white patches; (*b*) superficial veins.

middle of the channels between the dapples, and therefore are not so conspicuously coincident with the network. When, however, the pigmentary network is fine, it lies over and perfectly coincides with the venous network on the Horse's flank.

If you observe a dark-grey Horse and a white-grey Horse, you will see that in the former the white spots are disappearing, and mingling more and more with the dark-grey of the ground-colour, to form a uniform grizzly-grey colour, as in the upper Horse of Fig. 38; while in the latter the dark reticulations are lessening more and more, and restricted to mere vestiges, mainly on the shoulder and haunch, as in the lower Horse of the same figure. In many cases the white-grey Horse has nothing but the reticulations of the superficial veins to mark the places of former pigment reticulations, these having wholly disappeared, and having been replaced by a *uniform white colour*.

FIG. 41.—(*a*) and (*b*) are from the right hind-legs of two different Horses; (*c*) is from a light-bay Cart-horse (Whit Monday Show); (*d*) is a fern-like dappling over the superficial vein of the fore-leg of a dark-brown Horse (Whit Monday Show).

Certain well-dappled grey Horses have a very peculiar mark, like a fern-frond, which is pretty constant on the hind-legs, on the fleshy part between the heel and *knee*,[1] which is plainly visible in Fig. 36. This also coincides with a similarly disposed venous ramification on that particular part of the Horse's hind-leg. In the accompanying diagram (Fig. 41) are given three branching veins

[1] *Anatomical* heel and knee are here meant, and not the veterinary terms.

DAPPLED AND STRIPED HORSES

which correspond with this curious fern-like mark, and also a bay-coloured fern mark from the fore-leg of a brown Horse. The latter would perhaps correspond to those interesting marks on the fore-leg of Fig. 36.

The fern-like mark on the hind-leg of the grey Horse, when present, is always situated in the same place, viz., on the fleshy part of the hind-leg, in front of and a little above the hock.

On the legs of Horses the larger superficial veins have usually a somewhat transverse disposition, while on the flanks they are reticulate.

That the superficial venous distribution of the Horse has something to do with the pigmentation of its skin, I have not much doubt; but what that 'something' exactly is I am unable to say. Perhaps I ought to say the *nerves* of the veins have something to do with the pigmentation of the skin.

I have ransacked all kinds of works on the Horse in search of its general superficial venation, but have not found such a thing. Indeed, Professor M'Fadyean has told me that there has been yet no such publication.

The following, however, may be interesting to the student of the physiology of animal markings.

Nine months after I had written out my ideas on the origin of the markings of animals, and after I had received the photographs of a dappled Horse from Mr. Stanborough, of Bexhill-on-the-Sea, I read an article in the *Nineteenth Century* of April 1893, by Prince Krapotkin, on 'Recent Science.'

On p. 687 he writes thus:—

'Franz Werner's researches upon the colouring of Snakes, recently embodied in a separate work, show that the temporary and irregular spots which appear in Fishes and Frogs under the influence of artificial irritations are of the same character, and have

74   STUDIES IN THE EVOLUTION OF ANIMALS

the same origin, as the also temporary and irregular spots which appear in other Fishes, as well as in several Tritons and many Gekonides, without the interference of man. Some of the provoked changes of colour do not entirely vanish after the irritation is over, and they belong to the same category as the spots which appear in many animals in youth and disappear with growing age. Moreover, it is maintained that a series of slow gradations may be established between the irregular spots, the spots arranged in rays, and finally the stripes such as we see them in higher mammals like the Zebra or the Tiger; and if these generalisations prove to be correct, we shall thus have an unbroken series from the temporary spots provoked by light or electricity to the permanent markings of animals.'

I do not doubt that the pigments on the skins of animals are at one end of the telegraphic wires (the nerves) which connect them with the nerve-centres. Minute atomic changes, which, through age and other causes, occur in the nerve-centres, influence electrically the pigmentation of the skin; but what we have to search for is *why* all this is so. This investigation I have left for another place.

Let us now study a little in detail the dapplings of the grey Horse.

In Fig. 36, at the joining of the shoulder and trunk, may be seen several groups, consisting of a small roundish spot surrounded by larger polygonal jag-edged spots, something like the enlarged outlines shown in Fig. 42 (*a*).

In Fig. 38 (lower Horse) similar groups of maculations can be made out on the Horse's flank just behind the shoulder. Then in Fig. 34 there are three well-marked similar groups placed in a line slanting towards the abdomen, as well as several others.

In Fig. 42 (*b*) I have given two Jaguar rosettes for comparison. Those of (*a*) and of (*b*) are very similar; but I shall show in another

DAPPLED AND STRIPED HORSES 75

place that the *components* of these Horse dapplings are themselves only a fusion of minor rosettes—which I take to be the true homologues of the Jaguar rosettes.

Making allowance for the evolutionary deviations which have certainly occurred during the descents of the Horse and the Jaguar from their remote ancestral common stock, I hope to show that their skin-markings also result from *common* ancestral markings, although those of the Horse are now much altered. But so is his skeleton altered; and his legs, by a superficial observer, would

FIG. 42.—(*a*) Two rosette-like groups from the shoulder of Fig. 36; (*b*) two rosette-like groups from the flank of a Jaguar, given for comparison; (*c*) faint rosette-like dappling from the neck of a whitish-grey Horse.

scarcely be considered as having any community of descent with those of the Jaguar.

The spotting of the Horse would appear to be a transient feature, changing with age, like that of certain ruminants; while in the Jaguar the spotting may be much more permanent, although I am not aware that anybody has made any accurate observations on the strict permanency or otherwise of the Leopard's and Jaguar's rosetting. Judging from the specimens of very young Leopard's skins that I have seen, and from the very young Jaguar and Cheetah skins in the Natural History Museum, I should be inclined to say that the rosetting of these animals is not strictly

76 STUDIES IN THE EVOLUTION OF ANIMALS

permanent, although, looked at superficially, much alteration may not be noticed during the life of the same animal after reaching the adult stage. We know that in the Lion, the Puma, and various other animals the spotting alters so much, that in the adult it is on the verge of total obliteration.

In the dappled Horse the spots certainly tend to coalesce into

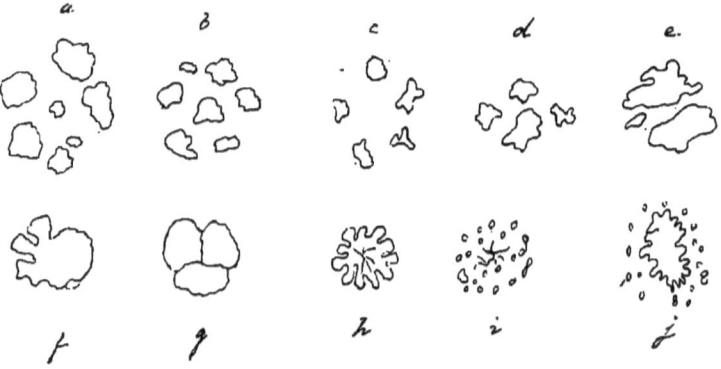

FIG. 43.—Various forms of rosettes seen on Horses:—(*a*) from flank of grey dappled Horse; (*b*) seen on both sides of a grey Horse, near the root of the tail; (*c*) from a Horse of the Great Northern Railway Company; (*d* and *e*) from dappled Horses; (*f* and *g*) from hindquarters of an omnibus Horse; (*h*, *i*, and *j*) Horse-rosettes breaking up.

frond-like patches, or to break up into star-like or other small spots, and even into minute specks, preparatory, it would seem, to total obliteration.

In Fig. 43 I have given several groups of spots—rosettes, we might call them—taken from various horses, which will elucidate what I have said above.

In this figure, it will be seen that in (*h*) the scolloped edges of the patch break up into an aggregation of a number of dots, as in

DAPPLED AND STRIPED HORSES

(i and j). A beginning of this may be seen on the groin of Fig. 38, that is, at the junction of the haunch with the trunk. When this breaking-up of the larger dapples extends all over the skin, we get, as I said, what is called the 'flea-bitten' horse. Now and again one sees a 'flea-bitten' horse which, on its haunches, has ordinary dapples. This would indicate that the minute spotting is only a *modification* of the dappled surface, and that the 'flea-bites' are probably the breaking-up of similar large spots.

A still further and more complete mixture of the two colours

FIG. 44.—Diagrammatic sketches of spots from various Horses :—(*a*) frequent disposition of dapples on flanks of Horses (right side); (*b*) occasional triangular spotting on the flanks of Horses (right side); (*c*) rare spotting on the flank and haunch of a Horse (left side).

of a dappled Horse produces the uniform roan, whether a strawberry or an ordinary roan.[1]

We have now to study another feature of the dappled Horse, which is only a modification of the ordinary confused mottling, if I may so call it.

As a rule, the flanks of the dappled Horse are simply reticulated, without any apparent order. But in certain dappled Horses the light-coloured spots on the flank occur in transverse rows, with the dark interspaces modified into broad lines or bands, both having the same slanting direction as the ribs, such as those shown in Fig. 44 (*a*). These rows of spots are usually squarish,

[1] These roans when newly clipped are often either bay-brown or black. Their heads usually do *not* roan.

78   STUDIES IN THE EVOLUTION OF ANIMALS

but in one case they were triangular, and seemed to fit into each other as in (*b*). This disposition of the flank spots and interspaces can, I think, be sufficiently made out in the dark Horse of Fig. 38.

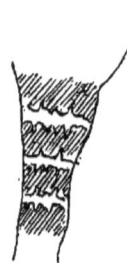

FIG. 45.—Rough sketch of fore-leg of a dappled grey Carthorse.

Moreover, in one particular Horse which I saw in the London streets, the spots in the front part of the haunch were also disposed in rows, which met the flank rows at the abdomen, as shown in (*c*).[1] Well, if in imagination you amalgamate all the white spots, you will have alternate light and dark transverse bands, and those of (*c*) would almost reproduce the haunch and flank stripes of the Zebra (Fig. 52).

This is not all, for I have not infrequently seen that, in the modified dapplings of grey Horses, on the upper part of the leg the white stars amalgamated into transverse irregular bands, while the dark reticulations also amalgamated and formed alternate bands. I have shown this feature roughly in Fig. 45. It is also very partially noticeable in Fig. 38 (upper Horse).

Further, on the neck of a light-bay pony at Bexhill-on-the-Sea, and also on a butcher's white pony in London, I have seen *faint* transverse stripes resembling those of the brindled Gnu. Then, on the upper ridge of the necks of grey dappled, and more especially brown dappled, Horses, the dark reticulations take the form of short bands. In a grey dappled Pony at the Horse Show of May 1893, the upper ridge of the neck was marked as I have

[1] The striped American Marmot (*Arctomys Hodii*) has each of its longitudinal stripes made up of a string of *white squares* on a dark ground (*Animal Kingdom of Cuvier*, by Griffith, vol. iii. p. 186).

DAPPLED AND STRIPED HORSES

roughly shown in Fig. 46 (*a*), and the hog-mane was divided into grey and white bands, just as it is in the Zebra. Then a brown Carriage-horse, reticulated with black, had the upper ridge of the neck as shown roughly in Fig. 46 (*b*), each space having a spot of a lighter colour than the ground-colour.

But the best example I have seen of stripes in the domestic Horse was on the 21st January 1894, in Piccadilly, opposite

FIG. 46.—(*a*) Hog-mane of a grey dappled Pony, Islington Horse Show, May 1893; (*b*) markings on the neck of a dappled brown Carriage-horse.

Berkeley Street. A milkman's cart had a Pony of a light-bay or dun colour. The Pony had broadish black stripes on his neck, shoulders, and flanks; he had also faint transverse stripes above his wrists and heels. A rough sketch of him is given in Fig. 47, which I took down at the time.

I have noticed that Ponies are oftener abnormally marked than larger-sized Horses. They appear to have more direct Ass and Zebra blood in them than the larger and more highly modified and artificially selected Horses.

Mr. Louis Robinson[1] suggests that the fossil Horse was about

[1] *North American Review*, April 1894, p. 483.

the size of the Shetland and Hungarian Ponies, and that these appear to be more nearly related to the original Horse, while the larger and more artificial breeds, like our breeds of Pigeons, Dogs, etc., are further removed from the original wild stock.

At the Whit Monday Cart-horse Show, I saw a dark-bay Horse with brindling on part of his flank, not unlike that of a brindled Dog.

FIG. 47.—Partially striped Pony of a dun colour.[1]

In this connection we should remember that, when the Zebra is crossed with the Ass, the striping, where it occurs, as is seen in the Natural History Museum, is not a Zebra *banding*, but a *brindling*, not unlike that of the Dog. (See Nos. 30 and 31, Appendix A.)

Then in dappled omnibus Horses, along their spine, I have

[1] A somewhat similar one is pictured on p. 322 of *Study of Animal Life*, by J. A. Thomson, referred to further on.

frequently seen marks like those shown roughly in Fig. 48, which are the marks we see along the spine of many Tigers, such as that given in Fig. 22.

The remarkable black dappling on the dorsal region of the white Horse given in Fig. 49 leaves little doubt that those blotches and spottings are broken-up Zebra bands.

All the cases I have quoted go to show that the dappling of Horses sometimes tends to dispose itself in a kind of striping or banding in some parts which is not unlike that of the Zebra. But independently of any theories of the genesis of striping in the Horse, Mr. Darwin, as a matter of fact, collected a large number of cases of Horses, which actually had stripes on their legs, their shoulders, and even on their faces. Mr. J. A. Thomson[1] gives the figure of a Devonshire Pony, from Darwin, with bands in certain parts which are evidently vestiges of a much more extensive ancestral banding.

FIG. 48. — Arrow-head marks along the spine of some grey dappled Horses.

Mr. Darwin[2] says, 'I have collected cases of leg and shoulder stripes in Horses of very different breeds in various countries from Britain to Eastern China, and from Norway in the North to the Malay Archipelago in the South. In all parts of the world, these stripes occur far oftenest in duns and Mouse-duns.'

These are the colours of wild Asses, and it would appear when these colours are reverted to, certain stripes, which belong to these wild animals, often reappear.

In the Zoological Gardens there is a specimen of the Asiatic

---

[1] *Study of Animal Life*, p. 322. Second edition.
[2] *Origin of Species* (1888), p. 200.

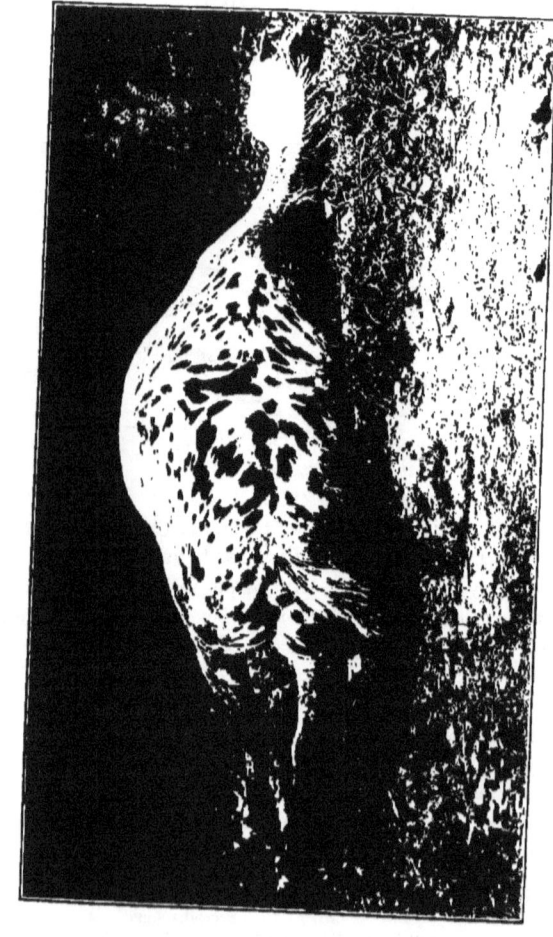

FIG. 49.—A blotched Horse, belonging to Mr. Douglas of Scone. A few hours before I had arranged to have him photographed, he broke his neck and died.

DAPPLED AND STRIPED HORSES 83

wild Ass (*Equus onager*). It is dun-coloured, with a broad spinal band like that of Burchell's Zebra, and faint transverse stripes above the hoofs.

FIG. 50.—Common Zebra, from a photograph by Mr. Gambier Bolton, F.Z.S.

The three Figs. 50-52 show with what perfection and completeness striping in the skin can be evolved in the genus *Equus*. They represent three variations of Zebra, viz., the common Zebra,

FIG. 51.—Burchell's Zebra, from a photograph by F. G. O. S. 10025.

FIG. 52.—*Equus Grevyi* (Grevy's Zebra), Somaliland, from a photograph by Mr. Gambier Bolton, F.Z.S., with obliteration of abdominal stripes (from a stuffed specimen).

FIG. 53.—Variation of Burchell's Zebra, with obliteration of stripes on legs and haunch, thus approaching the Quagga.

Burchell's Zebra, and Grevy's Zebra, a new addition from Somaliland,[1] which offers an astonishing example of animal striping.

At Mr. Rowland Ward's establishment I was shown the skin of what appears to be a variety of Burchell's Zebra. It had the broad striping and arrangement of striping characteristic of that species, but the intermediate paler bands were of a milk-white colour, and *without* any trace of those faint stripes between the black bands. I noted also that some of the broad black bands were made up of the fusion of two narrower bands.

There is another variety which is only partially banded, viz., the Quagga, given in Fig. 54. The article on the Quagga in the *Encyclopædia Britannica* (9th Edition) says: 'In length of ears and character of tail it more resembles the Horse than it does the Ass.... The colour of the head, neck, and upper parts of the body is reddish brown, irregularly banded and marked with dark brown stripes, stronger on the head and neck, and gradually becoming fainter, until lost behind the shoulder.' The haunch and legs have no stripes,[2] and it approaches the character of Burchell's Zebra of Fig. 53. Indeed, it has been frequently confounded with the latter by hunters. The Quagga was very common in South Africa, but now is very scarce.

In my opinion, the Zebra stripes, like those of the Tiger and other feline animals, owe their genesis to spots or rosettes or dapplings, which had become disposed in transverse rows, somewhat like those on the Cheetah of Fig. 10, and on the legs of the Jaguar of Fig. 4. Subsequently the rows of spots coalesced into beady bands, and ultimately became the sharp-edged bands we see in the Zebras.

[1] It has also been found on the shores of Lake Rudolf (narrative of Count Samuel Telekis' expedition in Equatorial Africa). Figure of the skin in *Proc. Zool. Soc. of London*, 1890, p. 413.
[2] There is a variation of this in the Science and Art Museum of Edinburgh.

FIG. 54.—Quagga, from Cassell's *Natural History*—(by kind permission of Messrs. Cassell & Company, Limited).

DAPPLED AND STRIPED HORSES

The Zebra has arrow-head marks on its shoulder. The domestic Cat of the British Museum, although partially spotted, has similar arrow-head marks in the same place, so has the Cat of Fig. 28, and the Tiger of Fig. 22, on the hind-legs. The rows of spots on the fore-legs of the Jaguar (Fig. 4) are disposed in arrow-head rows.

In the case of the spotted and striped Cats, as we have seen, it is comparatively easy to make out the derivation of the Tiger stripes from those of the Leopard rosettes; but it is not so easy to make out the derivation of the Zebra stripes from the spotting of the dappled Horse.

I have endeavoured to develop this idea as far as I could by means of existing examples which show the partial transition of dapples into bands. Further on, in Fig. 56 (a, b, c), I have shown that originally the rosetting of the Horses must have been closely assimilated to that of the Jaguar. The remainder must be left to the imagination of the reader, taking into consideration analogous transformations in the Leopards and Tigers.

We should note that the stripes in Zebras and Tigers, although they are transverse on the trunk, those on the legs are not a continuation of those on the trunk, as in the brindled Dog, but are also transverse, and correspond to the rows of leg spots of the Jaguar and Leopard. This order of things I have tried to account for as a result of inheritance from similar features in widely different ancestors.

With regard to the Ass, the only cases I have seen with vestiges of spots are those given in the Appendix C, Nos. 30 and 31. I have, however, seen several little Donkeys at the sea-side which had partial Zebra stripes.

Martin[1] says, 'While speaking of the white colour of some

[1] *History of the Horse*, by W. C. L. Martin (1845), p. 205.

breeds of the Ass, and the dappled markings of others, we may observe that a variety with Zebra-like stripes upon the limbs, to the very hoofs, is not unfrequently to be met with in our island and elsewhere, and sometimes even a double cross upon the shoulders is to be seen. To what cause the Zebra markings on the limbs (and we have seen them strongly painted in mules) are to be attributed, it is not easy to say. Is there, or has there been, a striped wild Ass indigenous in Asia,[1] or does this style of marking proclaim a cross at a remote date with some African species of the Zebra section?'

According to my view, all Horses and Asses and Zebras were originally spotted, or have descended from spotted ancestors. A large number among the domestic Horses have retained, or perhaps re-acquired, their spotting. From the spottings there resulted stripings. The Zebras have retained, or may have re-acquired, their striping. The wild and domestic Ass has got rid of both spots and stripes, but in some instances it has retained, or possibly re-acquired, *some vestiges* of striping. The same may be said of the self-coloured Horse, especially the dun-coloured Horse, the Kattiawar Horse, and the Mule. They frequently show vestiges of striping, and, as I said, the domestic Ass in two cases had vestiges of spotting or dappling on their flanks.

When a character ceases to be useful, it begins to disappear, and may become quite suppressed, while some other character more useful may take its place.

Fig. 55 is a picture of a 'Dhobi's' Donkey. I got it copied by a native artist. The original drawing is now in the Lucknow Museum. I have seen a somewhat similarly marked Ass in a picture which was hung in the refreshment room of the Army and Navy stores.

[1] We call the wild Ass an *Ass*, and the wild Zebra a *Zebra*, because they are differently marked, but in reality they are both *Asses*.

DAPPLED AND STRIPED HORSES

The question might now arise—If the striping of the Zebra is only a modification of some sort of ancestral Horse-dappling, why have not the stripings of the Zebra any signs whatever of their component spots or patches?

FIG. 55.—Picture of a Donkey in the possession of an Indian washerman.

Well, why have some Horses *no trace whatever* of dappling? Of course the spotting disappeared like many other things that become extinct. This disappearance is one of the great features of evolution. Atrophy first and total suppression later on ; and in colouring first faintness and then total disappearance, if circumstances do not allow of a part, or a colour, continuing.

Moreover, many Tigers in their striping show no evidence of its having originated from spotting. Yet if one makes a close analysis of the striping of a number of Tigers and the spotting of a number of Leopards, he can hardly remain unconvinced that the striping of the Tiger, though *primâ facie* without traces of its origin, owes its genesis to ancestral *spotting*.

The reader may perhaps think that I have been drawing somewhat on my imagination, and seeing things which may not be visible to others. But in the colossal museum of the streets of London, if any one feels disposed to study this interesting subject, by directing his attention to it, and by *keeping his eyes open*, he can amply satisfy himself that what I have stated is founded on fact.

In this connection it may be instructive to tell a little story. On one Sunday I accompanied to the Zoological Gardens a lady who was a Fellow of that Society. In going through the Lion and Tiger department, I pointed out how distinctly the Leopard spots could be seen on the legs and abdomen of the Lion.

She said, 'Do you know, I have been here hundreds of times, and have never seen those spots before!'

It is impossible to find in the same Horse all the features I have been discussing; one may be found in one, and another in another, because, like clouds, the dapplings shift; but one has only to put two and two together, and invoke the aid of the 'law of probability,' to come to conclusions similar to my own.

I have not, however, exhausted all the facts that can be adduced in support of my contention, viz., that the dapplings of the Horse are only greatly modified maculations of a nature originally not unlike those of the Jaguar, and that the striping of the Tiger and Zebra have originated from the stringing together—neck-

DAPPLED AND STRIPED HORSES

lace fashion—of spots, like those of Fig. 38 (upper Horse), which by further fusion were modified into clean-margined stripes. The similarity between the stripes of the Tiger and the Zebra can be seen at a glance; but the similarity between the spotting of the

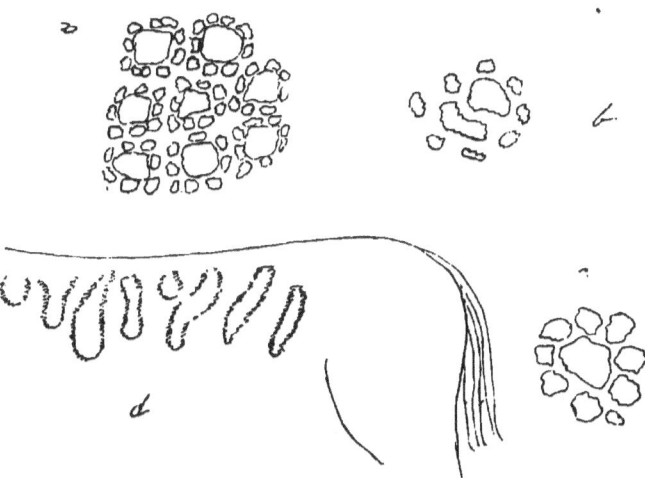

FIG. 56.—(a) Rosettes from the flank of an omnibus Horse of a strawberry roan colour (reduced); (b) rosette from the groin of a highly dappled grey Horse; (c) one of many rosettes on the flank of a dappled grey cart Horse; (d) marks on the flank of a brown Horse.

Jaguar and that of the Horse is not so easily traced, and has to be evolved with the aid of the imagination out of the variations presented by hundreds of specimens.

Nevertheless in Fig. 56 (a) I am able to give *bona fide* rosettes from the flank of a newly-clipped strawberry roan Horse. They are the marks I had long been searching for, under the conviction that, if my theory were right, I would discover them somewhere. They are rarely seen, because in the Horse the components of each rosette have fused into a continuous maculation, like those of Fig. 34. (*b*) Is a solitary rosette with its central space split into two portions ; (*c*) is one of many similar rosettes on a Horse ; and (*d*) were marks of a slightly different shade of colour from the general coloration of the Horse. Those of (*a*) are almost exactly like my restoration of the Jaguar rosettes of Fig. 70.

I do not, of course, mean that the dappled Horse will ever turn into a Zebra ; or the rosetted Jaguar into a Tiger. But I do think it quite conceivable that the striping of the Tiger evolved out of rosetting like that of the Jaguar ; and that both the Zebra and the dappled Horse may have evolved out of an animal with spotting not unlike that of the Jaguar. As to the relation of the Horse-dappling to the Zebra-striping, it only requires that one should keep his eyes open in the streets to see on the neck of dappled Horses, especially of Ponies, a decided tendency to Zebra-striping, which is often continued into the Horse's mane.

As I said, all experts agree that the Horse is born self-coloured,[1] and whatever the cause may be, if this state of things continues, the Horse will be a self-coloured animal ; otherwise he will be more or less dappled. The variations and gradations in the appearance and disappearance of spotting in animals are infinite. Some acquire spots at different ages, others lose them at different ages. Just as the embryo of the higher animals in evolving simulates the different stages of form through which, in its race history, the animal ascended, so the spotting and marking may give indication of the ancestral markings through which that animal in its evolution may have passed.

[1] Like the small Jaguar cub in the Natural History Museum—case 13.

DAPPLED RUMINANTS

There cannot be much doubt that the *Kiang*, or wild Ass, originated from a Zebra-marked ancestor. In the *Origin of Species*, p. 199, it is stated, on the authority of Colonel Poole, 'that the foals of this species are generally striped on the legs, and faintly on the shoulder.' In the adult they totally disappeared excepting the spinal stripe. So have the markings disappeared from the hind quarters and legs of the Quagga (Fig. 54).

Now let us turn for a moment from the Horse, the coloration of which I think I have sufficiently discussed, to another set of animals, viz., Giraffes and Bulls. From Fig. 57 it will be seen that the large blotches of the Giraffe are separated from each other by a lighter ground. These blotches may be nothing more than a *confluence* of a number of rosettes, such as we see on the shoulder of Fig. 34, or those on the fore-leg of Fig. 36. Indeed, on the shoulder of one of these figured Giraffes (left one), there are two blotches which seem distinctly to be made up of several smaller blotches.

Cows and Bulls are rarely seen with any vestigial marks of ancestral spotting or rosetting; but the Bull of Fig. 57, on its flank, shows distinct spotting not unlike that on the flank of the Horse in Fig. 37. Then I think no one will say that the striking rosetting on the flank of the Zebu of Fig. 58 is not almost identical with that on the shoulder and fore-leg of the Horse in Fig. 36.

The very extraordinary marks of this Zebu, so uncommon among domestic cattle, leave no doubt in my mind that the ruminants and the Horse are more closely allied than may have been supposed. Faint similar marks on the hump, abdomen, and haunches, lead to the presumption that in some ancestral ruminant the *whole skin* was covered with similar rosettes. Now domestic cattle are either blotched, piebald, or self-coloured; and Deer and Antelopes are either spotted, striped, or self-coloured.

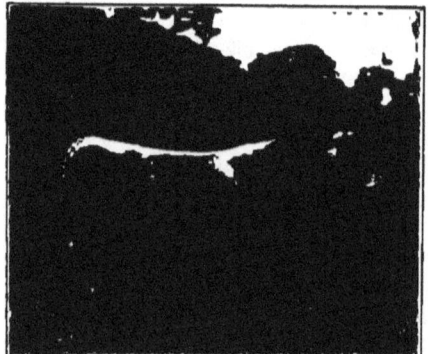

FIG. 57.—Giraffes and Bull, all three from photographs by Dixon & Son.

DAPPLED RUMINANTS 97

The Giraffe and the Zebu are the only ruminants, the one in a state of nature, and the other domesticated, which continue to reproduce distinct vestiges of ancestral skin markings, allied to those of the

FIG. 58.—Zebu, from a photograph, F. G. O. S., 10004.

Horse and the Jaguar. Indeed, if the rosettes on the flanks of the Jaguar skin of Fig. 4 had their enclosed spaces filled up they would become *Giraffe blotches*, and if we trim the rosettes of the Zebu and fill in their scolloped margins, we again reproduce the blotches of the Giraffe. In the Quagga we have seen that the stripes disappeared from the hind quarters and legs; so in the Giraffe we see the blotches disappearing from the under surface and legs.

Recently a new Giraffe has been discovered in Somaliland, and a description of it appeared in the *Saturday Review* of 21st July 1894, p. 72. It is distinguished from the South African Giraffe 'by a complete and whole body colouring of rich bright chestnut (I suppose the chestnut colour of the horse), scarcely separable by very fine,[1] almost invisible lines of creamy white of hexagonal shape;' while 'in its South African cousin the markings are widely and clearly defined,' as seen in Fig. 57.

Hunters in South Africa have often described the Giraffe as of a chestnut colour. This colour is common to the Horse and the Ox. This 'new Giraffe' does not appear to differ from the 'old Giraffe' any more than one grey dappled Horse differs from another grey dappled Horse. Compare different parts of the Horse in Fig. 34. In parts the reticulations are fine, in others they are broad, and in Fig. 37 they are still broader.

[1] At Rowland Ward's I was told that the divisions between the polygonal blotches were about one inch broad.

MEANING OF THE JAGUAR AND
LEOPARD ROSETTES
AND OF THE MARKINGS OF OTHER MAMMALS

'But, as soon as the attempt is made to think out the process in detail, we recognise that here, too, we know nothing thoroughly, and that it would be uncommonly easy for any one who wished to assign the processes of natural selection altogether to the realm of phantasy to emphasise this view : for *it is really very difficult to imagine this process of natural selection in its details ;* and to this day it is impossible to demonstrate it in any one point.'

The All-sufficiency of Natural Selection, by Professor AUG. WEISMANN, *Contemporary Review*, September 1893, p. 322.

PART III

MEANING OF THE JAGUAR AND LEOPARD ROSETTES AND OF THE MARKINGS OF OTHER MAMMALS

THE groups of spots, or rosettes as they are called, which these two animals present, are so characteristic that they must have some meaning, which hitherto, as far as I am aware, has not been worked out, or perhaps not worked out according to what, in my opinion, is the true meaning of the rosettes. What can that meaning be?

The Leopard spots, as we have seen, are merely contractions of those of the Jaguar; and may be the Cheetah spots are still further contractions and modifications of those of the Jaguar.

The larger rosettes of the Jaguar consist, as we have seen, of a certain space enclosed by a polygonal ring of spots (Fig. 4). The space enclosed has either one or several still smaller spots,[1] and sometimes none; and its colour is often different from the general ground colour.

In Fig. 59 I have given a number of variations of Jaguar and Leopard rosettes, and also some groups of Cheetah markings.

It is perfectly clear to me that in Nos. 10 to 13, and others, a fusion of some of the ring-spots has taken place, and a larger irregular patch has been the result; that in Nos. 3, 4, 5, 7 and 14, a

[1] Ordinary Leopard skins sometimes show a few rosettes with specks in the enclosed space.

fusion of the whole ring-spots has occurred; and that in Nos. 1 and 2 a still further fusion and contraction has resulted, *obliterating* the

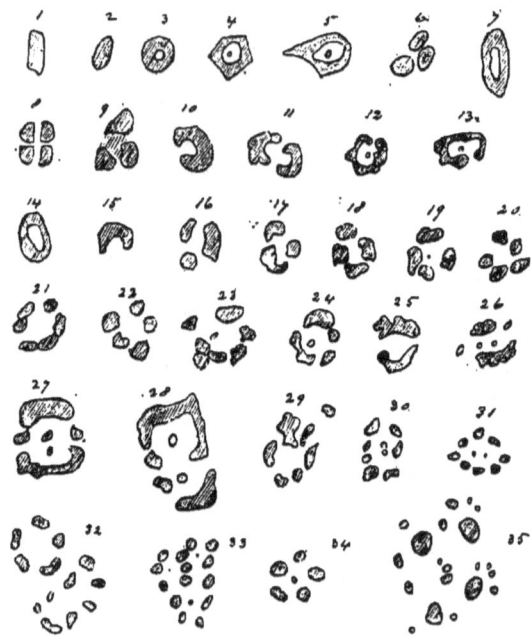

FIG. 59.—Nos. 5, 8, 9, 12, 13, 19, 26, and 30-32 are from Jaguar skins; Nos. 33-35 are from Cheetah skins; all the rest are from ordinary Leopard skins.

enclosed space. Several of the rosettes indicate transition stages towards partial or complete fusion. On the other hand, Nos. 26, 29, 30 and 31 would indicate a dissociation and tendency to scattering of the component spots of rosettes.

MEANING OF JAGUAR AND LEOPARD ROSETTES 103

The solid rosettes, Nos. 1 and 2, and other forms of solid rosettes, are to be found mainly along the spine and on the abdominal surface; while the dissociated rosettes shown in No. 32 are from the shoulder of the Jaguar. For reasons to be seen elsewhere I would consider No. 31 as the typical Jaguar rosette from which all the other modifications have resulted. The fused ring, when it occurs, may be either circular, polygonal, oval, or beaked, as seen in Nos. 3, 4, 5, and 7. It will be seen that the enclosed space, which is often of a *different shade*[1] from that of the general ground colour, sometimes contains one or more minute spots or specks, as in Nos. 27, 30, and 31.

Then No. 33 is a group from the flank of a Cheetah in the Natural History Museum; No. 34 is from a similar animal in the Zoological Gardens, and No. 35 is a group from the haunch of a Cheetah in the Science and Art Museum of Edinburgh.

These rosettes have been taken from various skins, but I think most of them can be matched from the skins of Figs. 4-7.

If the reader would take a pencil and draw outlines of the different variations in the rosettes of the skins given, he would be astonished with their variableness. Stretching of certain parts of the skin might perhaps, as I said, partly account for dissociation, and pressure and contraction might cause fusion, but I do not pretend to explain how all this occurs. Some idea of the process may perhaps be got by stating that in the embryo, while all the parts are semi-fluid, the pigment cells attract each other and fuse into bigger drops, just as minute discs of oil floating on water would now and again fuse into bigger discs if they came in contact, or would form groups of discs huddled together, and separated only by a capillary film of water.

This obviously would not account for the fact that, along the

[1] Note this.

spine, and on the abdomen, the rosettes coalesce into solid blotches, while on the flanks and shoulders they do not. This fusion and dispersion are evidently under the control of the nerve-centres, where modifications originate, and are only made appreciable by modified markings on the skin.

But, really, why the rosettes remain as rosettes on the flanks, and why they are dispersed on the shoulders of the Jaguar, while they coalesce and form solid blotches along the spine and other parts, I am unable to make out.

Prof. H. B. Orr[1] says: 'Any single nervous reaction, in a connected and often repeated series, does not disappear the instant its original stimulus is withdrawn. It will last for some time afterwards under the influence of association with the rest of the series. Eventually, however, it must disappear, sooner or later, according to the firmness of the association.'

In the Jaguar and Leopard it would appear that the nervous action is *very late* in disappearing, although they have been running on parallel lines with Lions and Pumas for thousands, and, may be, millions of years. They afford us an example of the *persistence* of influence, although the stimulus may have been withdrawn ages ago, while in Lions, Pumas, and others the markings have almost wholly disappeared.

It has been stated [2] that, at the Marine Laboratory of Plymouth, experiments on the under surface of flat fishes have been made by means of mirrors reflecting light from below. The under surface of the fish first became spotted, then the spots amalgamated, and finally the entire under-surface became dark. Photographs of the different stages have been taken.

According to Prof. Lodge, light is electricity, and experiments

[1] *A Theory of Development and Heredity*, p. 199.
[2] *Chambers's Journal* for August 26th, 1893, p. 541.

MEANING OF JAGUAR AND LEOPARD ROSETTES 105

may yet teach us a great deal about animal coloration and markings. The markings of the Jaguar, however, I think are due, not to the action of light, but to a totally different cause, as I hope to show.

Mr. Poulton[1] considers it proved that the white hair of certain Arctic animals is caused by the low temperature of the winter. Quoting from Mr. Welch, he says that in the American Hare, on the approach of winter, the dark hairs of the old coat turn white, commencing either at the tip, the middle, or the base of the hair, and that in addition there is a new winter coat coming up of an entirely white colour. And at p. 103, Mr. Poulton says that the young of the Arctic Hare are born grey, and turn white at their first winter.

All this does not affect the fact that in dappled grey Horses the spots are white, in dappled brown and dun Horses the spots are lighter than the ground colour, in roans they are black, and in strawberry roans they are bay or brown. Moreover, I have seen a grey Horse with tan-coloured spots.[2] So with many other animals that might be quoted. It is clear that in these cases temperature can have nothing to do with the difference in coloration, and we have to search for other causes to account for the coloration of certain animals, and more especially for those extraordinary markings we see in the Jaguar, the Leopard, the Cheetah, the Tiger, the Zebra, and so many other mammals.

A study of the rosettes in Fig. 59, and many others that can be seen on skins, makes it evident to me that, whatever may be the initial atomic nervous influence which produces these variations, they result in two sets of modifications, (*a*) a *fusion* of some or of all the encircling spots, and often in an obliteration of the enclosed space; or (*b*) a *dissociation* of the encircling spots, and perhaps even a scattering of the enclosed specks.

[1] *Colours of Animals*, p. 95.   [2] See Appendix C.

Between these two extremes, numerous intermediate variations of the rosettes occur, as the reader can see for himself.

I particularly dwell on the markings of the Jaguar, because they appear to me to be the typical markings, out of which many others we see in various animals have been elaborated or evolved.

It would be idle to suppose that *each* separate group of spotlets, forming a rosette, was made as it is by a process of natural selection. Such a notion would be wholly untenable. All the rosettes have a family resemblance, and therefore must have had a common origin. Natural selection may have had something to do with maintaining the *general* character of the spotted skin, but it could not possibly have had anything to do with moulding the outlines of each individual group of spotlets.

We must therefore look for a cause or causes of these markings which, at the same time that they allowed them to vary, stamped them with the features of a common parentage. If the Leopards have *retained* (not acquired) their very peculiar markings, so different from those of the generality of mammals, it would seem to follow that these animals emerged out of some more ancient mould *with* these peculiar marks. Their common family resemblance would point to their being modifications of a still older form of rosette, viz., that shown in No. 31, Fig. 59. How then did this more ancient form of rosette originate?—by natural selection or what?

Before I attempt to reply to this question in my own way, I would like to show how others have looked at it.

Dr. Wallace, in *Darwinism*, p. 288, says that Mr. Tylor[1] called attention to an important principle which underlies the various patterns or ornamental markings of animals, viz., that diversified coloration follows the chief lines of structure, and changes at

[1] *Coloration in Animals and Plants.*

MEANING OF JAGUAR AND LEOPARD ROSETTES 107

points, such as the joints, where function changes. Mr. Tylor says: 'If we take highly decorated species—that is, marked by alternate dark or light bands or spots, such as the Zebra, some Deer, or the carnivora, we find, first, that the region of the spinal column is marked by a dark stripe;[1] secondly, that the regions of the appendages or limbs are differently marked; thirdly, that the flanks are striped or spotted, along or between the regions of the lines of the ribs;[2] fourthly, that the shoulder and hip regions are marked by curved lines; fifthly, that the pattern and the direction of the lines or spots change at the head, neck, and every joint of the limbs; and lastly, that the tips of the ears, nose, tail and feet, and the eye, are emphasised in colour. In spotted animals the greatest length of the spot is generally in the direction of the largest development of the skeleton.'

Then at p. 289, still quoting from *Coloration in Animals and Plants*, Dr. Wallace says that 'Mr. Tylor was of opinion that the primitive form of ornamentation consisted of spots, the confluence of these in certain directions forming lines or bands; and these again sometimes coalescing into blotches, or into more or less uniform tints covering a large portion of the surface of the body. The young Lion and Tiger are both spotted; and in the Java Hog (*Sus vittatus*) very young animals are banded, but have spots over the shoulders and thighs. These spots run into stripes as the animal grows older; then the stripes expand, and at last, meeting together, the adult animal becomes of a uniform brown colour.'

And on p. 290, Dr. Wallace says: 'So many of the species of Deer are spotted, when young, that Darwin concludes the ancestral form from which all Deer are derived must have been spotted.

---

[1] Sometimes it is *white*, as in certain Antelopes, and also in the Kerry breed of cattle.
[2] Undoubtedly the markings often *cross* the ribs, as in the Ocelots and the *Viverridæ*, the Paca, etc.

Pigs and Tapirs are banded when young; an imported young specimen of *Tapirus Bairdi* was covered with white spots in longitudinal rows, here and there forming short stripes. Even the Horse, which Darwin supposes to be descended from a striped animal, is often spotted, as in the dappled Horses; and a great number show a tendency to spottiness, especially on the haunches.'

Ocelli may also be developed from spots or from bars, as pointed out by Mr. Darwin.

On p. 290, Dr. Wallace also mentions that in certain diseases the pigment is destroyed along the course of a nerve and its branches.

In this connection Mr. Darwin says:[1] 'Three accounts have been published in Eastern Prussia, of white and white-spotted Horses being greatly injured by eating mildewed and honeydewed vetches, every spot of skin bearing white hairs becoming inflamed and gangrenous.'

Other similar cases are quoted in other animals.

'We thus see that not only do those parts of the skin which bear white hair differ in a remarkable manner from those bearing hair of any other colour, but that some great constitutional difference must be correlated with the colour of the hair; for in the above mentioned cases, vegetable poisons caused fever, swelling of the head, as well as other symptoms, and even death, to all the white or white-spotted animals.'

Some great constitutional difference must be correlated with the colour of the hair. Undoubtedly; and the starting-point would seem to be the electricity of the nerve-centre cells, corresponding to the white parts of the skin, acted on electrically by the poison in the blood circulating among those central cells.

No one who has thought on this subject can doubt that the

[1] *Animals and Plants under Domestication*, vol. ii. p. 331, second edition.

MEANING OF JAGUAR AND LEOPARD ROSETTES

action of the nervous system influences the pigmentation of the skin, and the distribution of the pigments; but the nerves themselves seem to act only as *conductors* of the changes and influences which lie *deeper* in the cells of the *nerve-centres*. In electro-plating the wires conduct the influence of the action in the battery, and the metal is separated from the solution, and is deposited on the plate, or is dissolved off again, and returned to the solution according to circumstances. So in the pigmentation of the skin, the influence of the nerve-centres, acting through the nerves, puts on or takes off pigment, and modifies it according to circumstances.

Now the action of the nerve-centres depends on the blood, and the blood depends on food and physical surroundings. But in addition to all these, *heredity* undoubtedly plays a great part in all these phenomena. And in order to answer the question I put, we have to go back to features of still more ancient animals, which are now wholly extinct, viz., of *that large class of animals which carried armour plates on their skin*, as part of what is called an exoskeleton.

In remote times there were numerous animals protected by plate-armour; some of these have continued as survivals up to our time—such as Armadillos, Turtles, Crocodiles, Sturgeons, and many other curious fishes. Indeed, if we go lower down in the scale of life, we find that the exoskeleton is the only solid part of the animal structure, such as we see it in Crabs, Lobsters, and other *Arthropoda*.

Among armour-plated animals, now extinct, were several known species of Glyptodonts, like that shown in Fig. 60, the fossil original of which is in the Natural History Museum.

The armour of these strange animals consisted of either circular or many-sided plates, encircled by a rim of smaller polygonal

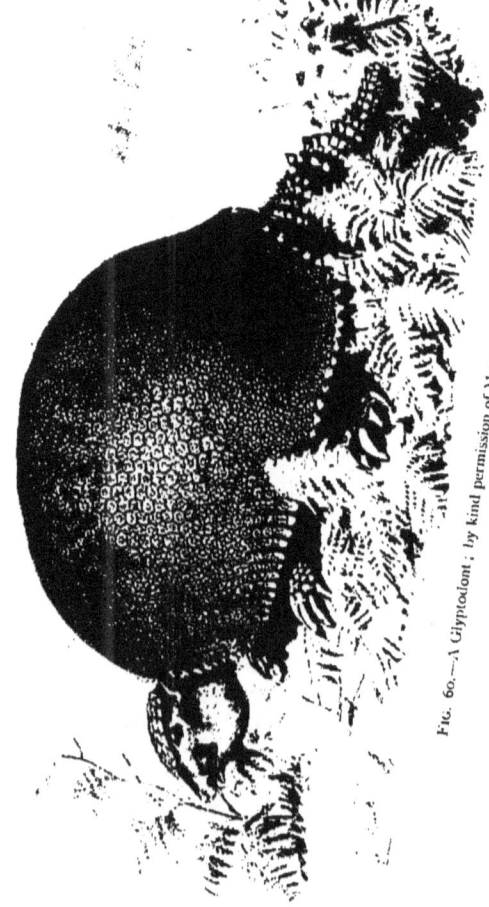

FIG. 60.—A Glyptodont; by kind permission of Messrs. Chapman and Hall.

MEANING OF JAGUAR AND LEOPARD ROSETTES 111

platelets, like those of Fig. 61 (*a*), (*b*), (*c*), (*g*). The form of these *bone-rosettes* varied not only in different species, but also in the same individual.

In the Natural History Museum, besides the Great Glyptodon, there are fragments of the carapaces of other species, as shown in Fig. 62.

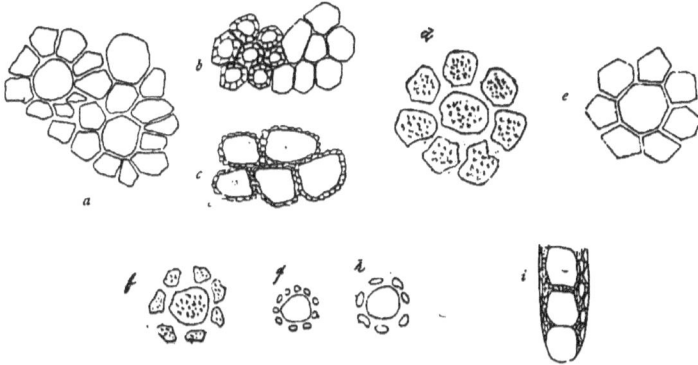

FIG. 61.—Bone-rosettes of various Glyptodonts (reduced): (*a*) from carapace of *Glypotodon asper*; (*b*) from front, and (*c*) from back of head-shield of ditto (Natural History Museum); (*d*) from *G. Clavipes*; (*e*) from *G. ornatus*; (*f, g* and *h*) from other Glyptodonts in the Royal College of Surgeons; (*i*) groups of tail plates of a Glyptodont (Natural History Museum).

Then in the museum of the Royal College of Surgeons there are remains of other forms of Glyptodonts, with bone-rosettes which perhaps are still more suggestive. In Fig. 61 (*d*), (*e*), (*f*), (*g*), (*h*), some of these rosettes are shown.

As these bone-rosettes[1] are taken from fossil remains, the edges of the component plates, and in some cases their surfaces,

[1] In the Natural History Museum, the plates of the Carapace of the Glyptodon and Armadillo are called 'ossification of the derm.'

112   STUDIES IN THE EVOLUTION OF ANIMALS

may have become partially eroded. Indeed, in some fragments, the platelets are almost wholly defaced, as shown in Fig. 62 (*b*). But I think that any unprejudiced evolutionist will not fail to see

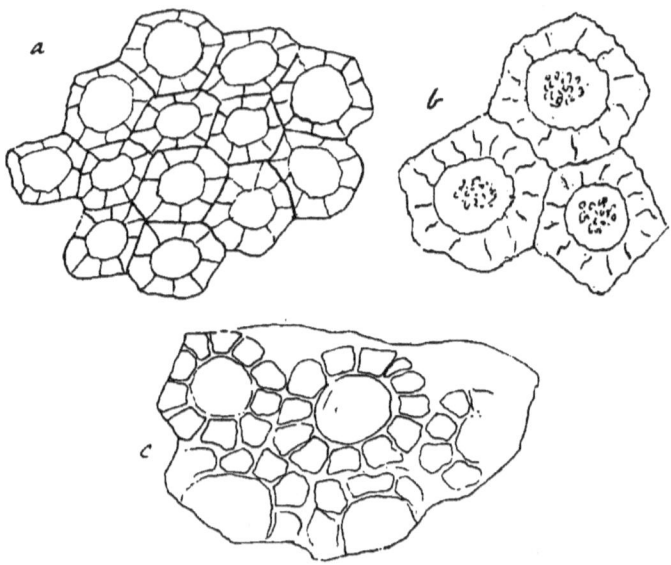

FIG. 62.—Bone-rosettes from other species: (*a*) from carapace of *Hoplophorus*; (*b*) from carapace of *H. Megeri*; (*c*) from fragment of a Hoplophorus (Natural History Museum).

in them the 'blocks,' so to speak, *from which the Jaguar got its imprints*!

Moreover, on the pelvic shield of *Polacanthus Foxii*, Fig. 63 (*a*), the remains of a Stegosaur in the Natural History Museum, and on the tail armour of a *Hoplophorus*, (*b*) we find armour-plates

MEANING OF JAGUAR AND LEOPARD ROSETTES 113

like those on Fig. 61 (*g*), which assimilate better perhaps with the typical Jaguar rosette shown in No. 31, Fig. 59.

Here I would note that the lumbar shield of *Tolypeutes tricincta*,

FIG. 63.—(*a*) Armour-plates of *Polacanthus Foxii* (Stegosauria) (reduced) ; (*b*) plates from the tail of a *Hoplophorus* (one-third natural size) from Fig. 1164, Nicholson & Lydekker's *Palæontology*, vol. ii. ; (*c*) plate from *Tolypeutes tricincta* ; (*d*) plate from *Priodontes Peba*.

an existing Armadillo, is composed of plates like those of Fig. 63 (*c*), and those of the Peba Armadillo (*Priodontes Peba*) are like those of (*d*) of the same figure, both which assimilate with the markings of the rosetted Cats.

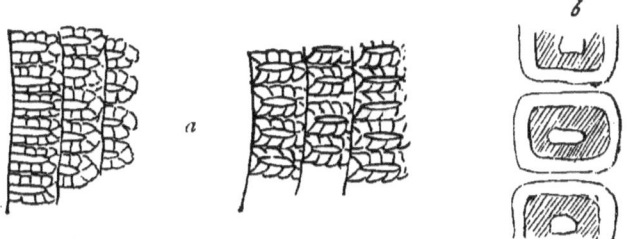

FIG. 64.—(*a*) Plates from shoulder and lumbar shields of *Tatusia Peba*, the seven-banded Armadillo ; (*b*) plates from the Great Armadillo, *Priodontes maximus*.—(Natural History Museum.)

Of course in some of the modern Armadillos the shield plates are much altered, but anybody can see that they consist of the same elements as shown in Fig. 64 (*a*). In (*b*) they are still more altered, and not impossibly the rim of each plate in this case is a *fusion* of the ring of platelets of others.

In Elliot's monograph of the *Felidæ* I met with a very curious assimilation between the disposition of plates on the forehead shield of an Armadillo, and the disposition of the coloration on the forehead of a domestic Cat. The latter forcibly recalls the former. Both are shown in the accompanying Fig. 65.

I have seen this forehead mark in other domestic Cats. Those specks which we see on the Jaguar skin, enclosed within the ring of

 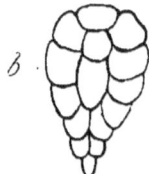

FIG. 65.—(*a*) Mark on the forehead of a domestic Cat (Elliot's *Felida*); the centre is brown, edged with black, and surrounded by a light ground; (*b*) forehead shield of an Armadillo, shown under the paw of *F. Yaguarundi* (Elliot's *Felidæ*).

spotlets, such as those of Nos. 26 to 28, Fig. 59, may just be the modified imprints of the little knobs we see on the hexagonal plates of *Tolypeutes*, a kind of Armadillo (Fig. 66 (*d*)). But recently Mr. Lydekker has possibly imparted to these inner specks a *special* interest. In No. 101, new series, of *Knowledge*, March 1894, Mr. Lydekker has described the club-tailed Glyptodont of Argentina. Each plate of this animal's carapace has several *holes* in it, which the writer supposes gave *passage to spines*. Whatever they may have given passage to, it strikes me very forcibly that the specks in the interior of the Jaguar rosettes,[1] or imprints, as I would call them, of ancestral carapacial plates, may possibly be the imprints

[1] I have seen rosettes on a Jaguar (Tring Museum) with as many as *six* specks.

MEANING OF JAGUAR AND LEOPARD ROSETTES 115

of holes *similar to those on the plates* of this curious *Dædicurus*, now in the La Plata Museum!

In past ages there were numerous carapaced animals even as low down as the fishes. Although I have been stating that the Jaguar and other Cats probably descended from Glyptodontoid ancestors, I do not mean that *all* the Cats descended from the *same* pair of ancestors, for it is obvious that some may have descended from one species, and others from others. Afterwards they may have intercrossed, just as they are able to do now. For instance, in the Cheetah, the marks are different from those of the Jaguar,

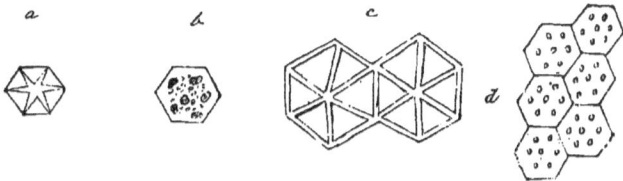

FIG. 66.—(*a*) and (*b*) are from Ostracions in the York Museum ; (*c*) from another Ostracion, an armoured fish ; (*d*) from pelvic shield of a Tolypeutes, an Armadillo (from Fig. 1161, vol. ii. of Nicholson and Lydekker's *Palæontology*).

and in the Horse we see transitions of marks which are like stars. Here are some patterns of plate armour for the reader to choose from (Fig. 66).

It may be asked, If other spotted animals have so much altered, why has not the Jaguar? Well, somewhat similar questions are often asked by women—' If we came from Monkeys, why do not Monkeys *now* turn into men and women?!' The answer is obvious to the mind of an evolutionist. The ancestral form, like every living thing, must either live or die. If it have endurance in its constitution, and is also suited to its surroundings, it will endure *side by side* with the variations which originated from it. In some cases it may endure *better*, and the variations may become extinct; while in other cases, the variations may endure better, and the parent form may become extinct.

Thus we see Monkeys themselves varied in a hundred ways, enduring side by side with man, who possibly originated from the Monkey-plane, and they may continue to endure until perhaps they become his rivals, and competitors for the consumption of his crops.

Professor Parker[1] says: 'Why such a form as the Glyptodon should have failed to keep his ground is a great mystery; nature seems to have built him, as Rome was built, for eternity. His exquisite little relative, the *Chlamydophorus*, scarcely larger than a Mole, has continued as yet to run out of danger, safe in his littleness; and many other kinds of low-brained mammals have, so to speak, the power to make themselves practically invisible.'

In another place I have ventured on a speculation which may account for the descendants of the Glyptodonts losing their carapace.

All these armour-plates of existing and extinct animals are *most suggestive*, and I consider them very important elements in the interpretation of the rosetting of the Jaguar.

After the armoured animals lost their skin-stiffening of lime deposit, the imprints of armour-plating must have become vastly more subject to modification than before, and so we have the endless forms of spotting and marking in mammals. After millions of generations that have been born since the Glyptodonts, and other Armadilloid animals, lost their bone-plates, it cannot be expected that the skin should retain the original outlines of plates unchanged. The great wonder is that the Jaguar has retained the features of its ancestral bone-plates *with so little modification*!

As an instance of the *slowness*, under favourable circumstances, with which long established characters are got rid of, both from the

[1] *Mammalian Descent*, p. 94.

MEANING OF JAGUAR AND LEOPARD ROSETTES 117

body and from its ruler the brain, the rosettes of the Leopards may be quoted. Who can tell how many millions of generations have elapsed since the ancestors of these animals and their congeners threw off their calcareous carapace, and adopted a masquerade of pigment-rosettes instead? Yet to this day this masquerading goes on in the Leopards and Jaguars, in their black varieties, as well as in the Snow Leopard. They are unable to shake off this pigment dress inherited from a calcareous carapace, although their habits and customs, their entire skeleton and teeth, have undergone great modifications.

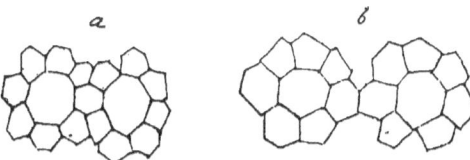

FIG. 67.—Bone-rosettes from the carapace of the Glyptodon, Natural History Museum, (a) partially fused rosettes from front of carapace, (b) distinct rosettes from side of carapace.

It must not be supposed, however, that bone-rosettes are not subject to fusion and other modifications. Fig. 67 (a) shows two bone-rosettes with a tendency to approximate and partially fuse; while (b) shows them quite distinct, and are given for comparison.

Hitherto I have traced the similarity of the Jaguar rosettes with the bone-rosettes of Glyptodontoid and Armadilloid animals which are also mammals. But *chi cerca trova*! Strange to say, this similarity can be traced much further down in the scale of life, and therefore much further back in time. Among the Chelonians we seem to find the identical mould from which the Glyptodon derived its bone-rosettes.

The Leathery Turtle (*Dermochelys coriacea*) has thin mosaically disposed plates on its carapace which at first sight seem a confusion of platelets without any order; but further investigation reveals on the right side,[1] below one of the longitudinal crests, the fact that several of them are disposed in distinct rosettes

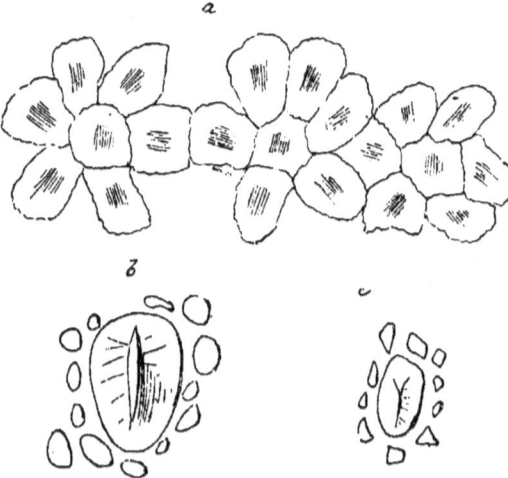

FIG. 68.—(a) Bone-rosettes on the carapace of the Leathery Turtle (*Dermochelys coriacea*); (b) crested plate of Nile Crocodile, encircled by small plates from right flank; (c) crested plate of Sturgeon (*Accipenser sturio*) seen between the larger plates on the spine and flank; all from the Natural History Museum.

exactly like those of the Glyptodon, as shown in Fig. 68 (a). Curiously enough, this Turtle is one of those whose carapace is *not* like that of other Turtles, fused with the spine and ribs. There is therefore some suspicion that it may be one of the ancestral forms of such mammals as the Glyptodonts, otherwise it would be strange

[1] Specimen in Natural History Museum.

MEANING OF JAGUAR AND LEOPARD ROSETTES 119

that the Leathery Turtle should exhibit the identical bone-rosettes of the Glyptodon. They are no doubt very much *thinner*, as if they may be either forming or disappearing; nevertheless, no one can doubt that they are the same things, and are produced by the same cause, whatever that may be. It is also noteworthy that two of the Turtle rosettes show a tendency to approximation and partial fusion, like those of Fig. 67. This community of features points to the probability of the Glyptodonts having emerged from some Turtle-like water-animal.

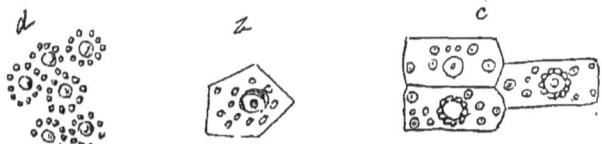

FIG. 69.—(*a*) Group of ornament tubercles on head-shield of *Encephalaspis Pagei* (Prof. E. R. Lankester's *Monograph of the Fishes of the Old Red Sandstone of Britain*, p. 49). The scales have the same character of markings; (*b*) ornament in polygonal plates tuberculated of *Hemicyclaspis Murchisoni*, p. 52; (*c*) Portion of test or shell of *Echinus gracilis* (*Zoology of the Invertebrata*, by A. E. Shipley p. 241).

Then the Nile Crocodile and the Sturgeon show in their armour forms of plating similar to the imprints of the Jaguar, as seen in Fig. 68 (*b*) and (*c*).

Lower and lower still in the scale of life we meet with the same curious pattern in its more primitive form, viz., that of a central larger plate surrounded by a ring of smaller plates or tubercles. I would not venture to suggest that the spotting of the Cheetah has any relationship with the plates on the test of the Echinus, or with those of armoured fishes; for, if I did, I might be laughed at, and therefore I simply give the markings on these animals in Fig. 69 so that the reader may judge for himself.

(a) Shows the ornamentation on the head-shields of *Encephalaspis Pagei*, an extinct Sturgeon-like fish; (b) shows a plate from *Hemicyclaspis Murchisoni*, an extinct animal like an Armadillo; while (c) are from the test of an Echinus.

Then, curiously enough, the Pearly Nautilus in the Natural History Museum, on its soft hood exhibits a similar pattern, although it does not seem to have there any calcareous deposit. The soft hood appears to act as an operculum when the cephalopod withdraws into its chamber, and therefore may at one time have been *armoured*.

It would seem preposterous to endeavour to carry on the relationship of the Jaguar and Cheetah markings to those on the armour of the Cephalaspidæ,[1] or the Echini. Yet if we are evolutionists, and if creation by the method of evolution be true, the Echinus and other low forms cannot be left out in the cold because such hints as I have made would outrage our feelings! We must look upon the markings of certain fishes and of the Echini as connected *somehow* with those of the higher land animals.

Evolutionists go even further than this, and admit that all animal forms on the earth were evolved from the minute pelagic forms, many of which are still in existence. The doctrine of evolution would teach us that all these phenomena are rather a matter of course than mere unexplainable anomalies.

It does not, I think, require much acumen to see a family resemblance between the figures of the Glyptodon's bone-rosettes and the spot-rosettes on the flanks of the Jaguar. Indeed, a glance at the back and flanks of Fig. 4 will readily suggest the impression of a carapace composed of plates not dissimilar to those of the Glyptodon. I do not say that the Jaguar descended from a Glyptodon, but I do say that this mammal descended from some extinct

[1] The Cephalaspidæ were Sturgeon-like ancient extinct fishes.

MEANING OF JAGUAR AND LEOPARD ROSETTES

animal with a Glyptodon*toid* carapace. I repeat that the bone-plate rosettes on Figs. 61 and 62 speak to us only too plainly. These armour-plates of extinct animals are to my mind the 'blocks' which gave the Jaguar skin the impressions of the groups of spots or rosettes, much modified in subsequent innumerable generations. In other words, the skin, on losing its hard calcareous plates, *retained somehow an impression of them, which modified the pigmentation, where the plates in its ancestors originally stood.* Or again, to put it more 'nervously,' the action of the nerve-centres which caused the deposition of calcareous matter in rosette-form, on the skin of the Glyptodonts, *continued* to act when there was an *insufficient* amount of calcareous matter in the blood *for this purpose*.[1] That is, in the nerve-centres a sort of *memory* of former plates remained, which expressed itself in *pigments* after the calcareous carapace had gone. This nervous action then resulted in the deposition of pigments of colours different from those of the general skin. In many mammals this nervous action dwindled into the deposition of simple spots; in others it fused them into lines and patches, and in some they were entirely obliterated, as in the adult Puma and others.

The intervals between the Jaguar and Leopard rosettes—now altered, as I said, through innumerable generations—would seem to indicate the *sutures* between the armour-plate rosettes of some ancestral Glyptodontoid animal.

Of course the skin of the Jaguar is elastic and mobile, and stretches readily to adapt itself to the growing animal and to the different variations in size which we see in the Cat tribe. Its elasticity would seem sufficient to account for the *broadening* of the network of sutures which we see everywhere between the Jaguar rosettes, and perhaps also for the dissociation of the spotlets which

---

[1] In the Natural History Museum there is an interesting series of skins of Lacertilia; some have fully ossified scales, others have only *vestiges* of ossification.

122 STUDIES IN THE EVOLUTION OF ANIMALS

compose the groups on its shoulders (Fig. 4); and not improbably the stretching of the skin might partially account for the characteristic spotting on the haunch of the Cheetah shown in Fig. 59, Nos. 33-35.

In Fig. 70 (*a*) I have endeavoured to restore some of the plates of the carapace of the Jaguar's immediate ancestor. Of course this restoration is imaginary, but it is suggested by five contiguous rosettes on the Jaguar's left flank, behind the shoulder (Fig. 4). I

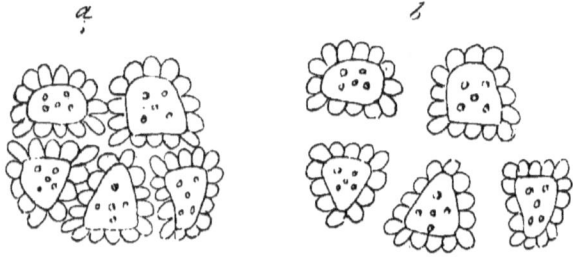

FIG. 70.—(*a*) Restoration of the rosettes of an imaginary Glyptodontoid ancestor of the Jaguar, with closely interlocking plates ; (*b*) The same in subsequent forms of mammals, when the plates may have become dissociated, with intervening flexible skin.

have shown the plate-rosettes as closely fitting in (*a*); but imagine them to have become somehow dissociated as in (*b*), from causes which I have already discussed, and you have intervening channels or commissures of flexible skin, such as we see in the Crocodiles, and also between the bands of the Armadillos. These skin commissures are now represented in the Jaguar by the paler ground colour reticulated between the rosettes. I have given the imagined plates a trapezoid shape, to make them similar to the Jaguar rosettes, but the shape of the latter may have become much distorted, as indeed we see it in various parts of the Jaguar skin itself. I have

MEANING OF JAGUAR AND LEOPARD ROSETTES 123

also given five little specks on each central plate, to assimilate them with some of the Jaguar rosettes, but there may have been many more, as in the plates of *Tolypeutes*.

All this imaginary restoration, however, was hardly necessary, as in Fig. 62 (*c*), which is part of the carapace of a Glyptodont, shows an almost *identical* arrangement of bone-rosettes.

Reference to Figs. 61 and 62 will show how various the number of small plates is which encircle the bigger central plate. In some cases the inference would be that the smaller plates have occurred to completely fill in vacant spaces left by the larger and more solid plates, and thus leave no part of the skin unprotected. Where great flexibility is needed, as between the bands of Armadillos, it is obvious that plates would be an encumbrance, and simple elastic skin preferable.

The dissociation of the plate-rosettes, with intervening elastic skin, would have admitted of *freer movements*—a feature of great importance in the struggle for existence—than would have been possible in the dish-cover solid carapace of a Glyptodont.

This dissociation of plate-rosettes is not simply a hypothesis, for we see it actually occurring on the abdomen of the Great Armadillo (*Priodontes maximus*),[1] which in that region has rows of *separate* rosettes, composed of minute plates, while the carapace is formed of bands of squarish plates, with intermediate skin commissures, like those shown in Fig. 64 (*b*).

It may perhaps be objected, that if the plate-rosettes could be dissociated bodily, the component smaller plates could also be dissociated. Just so, and that is what may have occurred in the ancestry from which the Cheetah has descended. Its plate-impressions, large and small, though probably much modified, are scattered all over its skin.

[1] Edinburgh Museum of Science and Art.

In snakes we see closely fitting scales, but in the Python, and in the expanded hood of the Cobra, the scales are scattered over the skin, with intervening spaces *without* scales. So that stretching of the skin may have had something to do with the scattering of the rosettes, whether as dissociated bone-plates, or dissociated pigment-spots.

FIG. 71.—The Hairy Armadillo (*Dasypus villosus*), Natural History Museum.

There is another feature in the Leopards which is worthy of notice. On their legs there are *transverse* rows of *small spots*. These, in my view, are impressions left by transverse rows of *small plates*, not unlike those which we see on the legs of an existing Armadillo, shown in Fig. 71.

We now begin to get a sort of 'moral conviction' that the transverse stripes on the legs of Tigers, and other Cats, are due to ancestral rows of spots like those on the legs of Leopards, and these in turn are due to the fact that their armoured ancestors had *scales* or plates there, *similarly disposed*.

In the Hairy Armadillo and others no doubt the leg-plates are in process of extinction. As the animals can roll themselves up in their banded carapace, their leg-plates have become superfluous, and are doomed to extinction.

Taking then everything into consideration, I think it will be found difficult to escape from the conclusion that the markings of the Leopards are *inherited* from ancestral plate-impressions of some extinct Glyptodontoid form, and have *not* been evolved by a process of natural selection.

It is a wonder to me that in the Jaguar so much likeness to ancestral bone-plates still remains as to enable us, through these hieroglyphics, to read the story of its descent. But for these

MEANING OF JAGUAR AND LEOPARD ROSETTES 125

Jaguars and Leopards, the ever-recurring variations which we see in living mammals would have prevented us from deciphering markings which are all but universal in mammals, but so strangely altered ; for who would have conjectured that the stripes of the Tiger had anything to do with an armour-plated ancestral carapace ?

The very fact of those great Armadillos having become extinct may have resulted precisely from the unstretchable nature of their bony skins. Their movements must have been hampered to a vast degree ; and the great difference between their edentate jaws and the jaws of the carnivora is enough to indicate to us the vast changes that must have occurred from the days of the Glyptodonts to those of the existing carnivora.

In another place I have shown that the markings of a large number of carnivora are traceable to modifications of the markings of the Jaguar. I have also shown that some Horses have rosettes (Fig. 56 (*a*)) not very dissimilar to those imaginary ones of Fig. 70. I take the dappling of Horses to mean a fusion of the ancestral Horse rosettes ; and the Zebra banding again as a modification of the ancestral dappling. Then the rosetting, spotting, and striping of the ruminants would be further modifications of the original Horse-like *rosetting*. In the 'old Giraffe' we seem to have a fusion of a number of rosettes to form the polygonal blotches, with *broad* commissures ; while in the 'new Giraffe' of Somaliland, the polygonal blotches are approximated and the commissures are not so broad.

Further on I have endeavoured to show cause for the thinning, disintegration, and the final disappearance of such massive calcareous carapaces like those of the huge extinct Glyptodonts from which all this picture rosetting, spotting, blotching, striping, and banding in existing animals, as I think, have come.

After the exoskeleton was got rid of, the internal skeleton went on evolving *independently* of the coloration of the skin, and that may be the reason why we find almost identical coloration or markings in mammals, with vast modifications in their endoskeleton.

What were at one time solid plates with holes or knobs, became later on *expressed* in simple pigments of different colours to the great advantage of the losers of such a stiff and unmanageable encumbrance.

If it be true that the rosetting of the Jaguar and Leopard originated as I have stated in the foregoing pages, the same theory will account for the markings of Horses and other mammals, including ruminants and more especially Oxen, as testified by the rosette markings of the *Zebu* (Fig. 58).

POSTSCRIPT

WHILE this book was passing through the press, I read in *Knowledge* of January 1895 a paper by Mr. R. Lydekker on the 'Spots and Stripes in Mammals,' wherein he gives Professor Eimer's and his own views of the origin of mammal coloration, as he does not seem to agree with Professor Eimer.

I am sorry to disagree with two great authorities on these matters, but Thomas Carlyle's teaching was this—say what you *think*, even if you are gibbeted for it.

I cannot make this subject intelligible to the reader without giving extracts of some length from Mr. Lydekker's paper; and perhaps it will be better if I take certain points of his paper one after the other, and try to discuss them, as the questions are exceptionally interesting.

On p. 3 he says :—

'These markings generally take the form either of longitudinal or transverse bands or spots, the latter being frequently arranged in more or less distinctly defined longitudinal lines, but " never " in transverse bands.'

POSTSCRIPT 127

I think I have shown that both in the Leopard and in the Horse spots *do* not infrequently arrange themselves in *transverse* lines or series.

P. 4. 'A similar state of things occurs among wild Pigs, and also in Tapirs, from which we are naturally led to infer that in this group of mammals, at least, a spotted or striped type of coloration is the "original" or generalised condition, while a uniformly coloured coat is an acquired or specialised feature, and we shall find that this will hold good for other groups.'

In my humble opinion—and I have tried to show why I hold this opinion—both the *striped type* of coloration and the uniformly coloured coat are *derived* from the spotted or rosetted types.

In writing of the Rodents, Mr. Lydekker says (p. 4) :—

'A survey of the collection of these animals in a good museum will show that, whether the patterns take the form of stripes or spots, the arrangement is invariably longitudinal, and "never" transverse.'

Now, it is impossible to separate the Rodents from other mammals, and, as I have already said, transverse rows of spots and rosettes are not uncommon in other mammals. All that we can infer is, that in the *existing* Rodents no species with transverse stripes or spots are found. Perhaps this is what Mr. Lydekker meant.

P. 4. 'By a splitting-up of a simple spot into a more or less complete ring of smaller ones, we have the rosette-like type of ornamentation, as exemplified in the Leopard, the Snow Leopard, and the Jaguar. In the two former, the ring encloses a uniform light area ; but in the latter the central area generally carries one or more dark spots.'

I confess I am unable to accept Mr. Lydekker's view of the origin of the rosette from a *splitting-up of a simple spot*, for reasons given in Part III. I am indebted to Mr. Lydekker himself for a hint of the *value* of those interesting 'one or more dark spots' in the central area of the Jaguar rosette.

In one of the numbers of *Knowledge* he described a strange Glyptodont (*Dædicurus*), with a club tail, in the La Plata Museum. The *holes* in each armour-plate of that singular animal are, in my opinion, the *equivalents* of the dark spots in the interior of the Jaguar rosettes, a remarkable example of which is to be seen in the Tring Museum.

Then, in the beautiful illustration of the Snow Leopard in Mr. Lydekker's paper under discussion, we seem to follow the genesis of the stripes on the tail *gradually* from the rosettes on the haunch of the animal. The rosettes are carried on to the base of the tail, and insensibly, by flattening and obliteration of the enclosed area, they are converted into stripes. So that we have the choice of *two* modes of genesis of stripes—either, as I have shown, from rows of rosettes, like the twin stripes of certain Tiger skins, or from single rosettes, as indicated above.

P. 4. 'A further development of the ring leads to the so-called clouded type, as displayed by the Oriental Clouded Leopard, the Marbled Cat, and the American Ocelot. Here the ring becomes enlarged into a large squarish or oblong area, enclosing an area of darker hue than the general ground-colour of the fur, and bordered by a narrow black line.'

In discussing the markings of the Ocelot, I have shown that their curious markings result from a *fusion* of many rosettes disposed in longitudinal order, and not from an *enlargement* of 'the ring.' In the Ocelots it is a *longitudinal* fusion of a *number* of rosettes; in the others it is a squarish patch, formed, as I think, by the fusion of a *group* of rosettes into one large island. Many modifications may have occurred after fusion, and in the Clouded Leopard we have perhaps a transition from this into the eventual uniformly coloured coat—a sort of breaking-up of the whole patch.

The curious part is that in some Ocelots the central dark spots of the rosettes are still retained, but they become arranged in line in the centre of the longitudinal band, while the *amalgamated rings* form a *black border* to the scolloped band.

P. 5. Professor Eimer, 'as the result of his investigations, laid down the following laws:—

'*Firstly*, the primitive type of coloration took the form of longitudinal stripes; *secondly*, these stripes broke up into spots, retaining in many cases a more or less distinct longitudinal arrangement; *thirdly*, the spots again coalesced, but this time into transverse stripes. And, further, all markings disappeared, so as to produce a uniform coloration of the coat.'

To this Mr. Lydekker says (p. 5):—

'There ought, if the theory were true in its entirety, to be a considerable number of longitudinally striped species among the lowest groups of all. . . .

Professor Eimer makes no distinction between light and dark markings . . . nevertheless we may provisionally consider light and dark stripes and light and dark spots as respectively equivalent to one another.'

I presume that by the 'lowest groups of all' Mr. Lydekker means the marsupials. Biologists look upon these animals as the most *primitive* types of mammals, and therefore the oldest. This, I should say, is a good reason for believing that they had, compared with others, *a far longer time to change in*, conformably with surroundings of sorts, and therefore it must be a wonder that we meet with *any* of them which are marked at all with either spots, stripes, or bands. Nevertheless some of the marked ones still survive!

This, in my humble opinion, is sufficient evidence that both they and the other mammals of superior organisation *came from the same marked stock*.

P. 5. 'The fact that the markings of young Pigs take the form of longitudinal stripes, whereas in the more specialised Deer, whether young or old, they are in the shape of spots arranged in more or less well-defined lines, is, as far as it goes, a confirmation of the theory that "spots" are newer than stripes.'

My own studies of this interesting question have led me to just the opposite conclusion—viz., that spots are older than stripes, and that rosettes are the *oldest* of all markings.

Perhaps too much stress has been laid on the word *specialised*. Although the skeleton may be highly specialised, it does not appear to me to *follow* that the exterior of the animal will march *pari passu* with its interior. Indeed we know that it does not. For the young and the adult of the Puma are both equally specialised, yet the former is spotted, while the latter is plain. It seems to me that the exterior of a mammal would be more *actively* influenced by its surroundings of sorts than its interior. And so it happens that we find, among so many differently specialised orders, *survivals* of ancient coloration, which occur *in spite* of the specialisation conformable with the order to which each may now belong.

With regard to the longitudinal stripe down the back, several views seem to suggest themselves as to its origin: (*a*) It may be an extreme contraction of a broadly dark back, such as that of some Squirrels; (*b*) It may be a fusion of the oblong black spots which we see along the spine of many Leopards. It may persist *longer*, for some reason, along the spine, when on the flanks, shoulders, and hind-quarters all markings have disappeared. The total obliteration of markings we begin to notice on the hind quarters and legs of the Quagga. Either the spinal line, or the leg stripes, or both, seem to persist when the body becomes plain. Why it should be so, if it is so, I do not know; (c) I have tried to account for this spinal dark line, as a picture-remnant of the dorsal ligament, which, in the little Pichiciago, binds the horny carapace to the spinal line.

This curious little Armadilloid mammal seems to be a survival of a *transition* stage between the bony-carapaced Glyptodonts and the hairy mammals, the former, as I think, having been the fathers and mothers, not only of carapaced mammals, but also of *all* hairy mammals; and I take it the Jaguar presents us with good evidence of all this.

But after all this dorsal line may perhaps be of little importance; for any one can see, in the Tiger skins of the furriers' shops, that, in certain specimens, some of the transverse stripes *merely meet* at the spine, while others, after meeting, are conjointly *prolonged* in the direction of the tail. What is curious is this: that, in some mammals, this dorsal line is *white*.

P. 5. 'Then again, in the ungulates we have the Zebra-Antelope, the Gnus, and the Zebras showing most strongly-marked transverse dark stripes; but we have no dark-spotted forms in the whole order except the Giraffes; while the only ones with dark longitudinal stripes are young Pigs. And it would thus appear that, although all the animals above mentioned are highly specialised species, these transverse stripes and dark blotches must have originated *de novo* quite independently of the groups in question.'

I must humbly beg to differ from all this. I have shown that Horses do not unfrequently have *dark spots*—even one little Donkey had them —and that, in the grey-dappled Horse, the white spots are sometimes arranged in transverse series of lines. These, by fusion, would give an alternation of white and dark stripes. The blotches of the Giraffe assimilate with the blotches of the Horse. Then the rosettes of the Zebu of this work are almost *identical* with those of certain dappled Horses.

As the coloration both of the Giraffe and the Zebra harmonises with their surroundings, Mr. Lydekker thinks—p. 6, that,

'It is incredible that both types should have been evolved, according to a rigid rule from animals marked by dark longitudinal stripes.'

That, is from the *primitive* markings of Professor Eimer's theory. Quite so, but, as I have shown, it is not so incredible that both the coloration of the Giraffe and of the Zebra, as well as those of the dappled Horse and of the Zebu, should have been evolved from, in my opinion, the oldest coloration of all, viz., the *rosettes*, similar to those of the Jaguar.

Mr. Lydekker in conclusion says that it is not improbable that there may be a certain substratum of truth in Professor Eimer's theory.

'What we may call the "longitudinal-spotted-transverse-uniform" theory of coloration, we submit that in its present guise it cannot adequately explain the whole evolution of "spots and stripes in mammals."'

Mr. Lydekker's paper on this subject in *Knowledge* is very well worth the attention of students of the coloration of mammals. It is, moreover, accompanied by the most beautiful illustration of the Snow Leopard that I have yet seen.

Now, may I be allowed to put forth in concluding this P.S. my theory of the whole question, as succinctly as I can—as Carlyle might have said,—the theory of a poor simple creature?

The Glyptodonts, or other armoured mammals of a similar nature, were the *originals* from which all existing mammals, including marsupials, descended.

The Jaguar, for some reason or other, has retained the most primitive type of coloration, due to the characters of the ancestral armour-*plates*— a sort of *picturation* of the carapace, after this had been wholly got rid of.

All other spotted mammals—whether longitudinally, or transversely, or diagonally—are *modifications* of those of the Jaguar.

Then the stripes—whether longitudinal, transverse, or diagonal—are fusions of lines of spots or of rosettes. This seems clear from the spotting of certain Cheetahs, certain Horses, and certain Tigers with twin-stripes.

The markings of the Ocelots, Clouded Leopards, and Marbled Cats are a fusion and *modification* of groups of rosettes, either longitudinally disposed, or grouped in irregular patches.

The piebalding of the Dog seems to be made up of fusions and agglomerations of spots, as may be seen in certain Dalmatian and other Dogs.

Then in the self-coloured mammals, it is evident there is, for some reason, a total obliteration of all special marking, although they now and again turn up as atavic marks, perhaps by some atomic disturbance of crossing in the nerve-centres.

The curious point is, that white, black, and tan colours are *interchangeable*—why, I do not know. This can be readily seen in the Fox-Terrier. The black patches of the one are almost identical with the tan patches of the next. The black and tan colours of the one become tan and white colours of the next, and so on. Even the tan spots over the eyes have been known to change into white spots. So that it is no wonder that we see black spots or stripes in one order of mammals changing into modified white spots, or modified white stripes, in another order of mammals.

The interchangeableness of these three colours, with variations in the shades of the tan, can be seen in many other mammals, both domesticated and wild.

Lastly, when all vestiges of rosetting, spotting, or striping have disappeared, as in self-coloured mammals, there may remain a vestige of the ancestral carapace *as a whole*, without any vestige of the separate plates. This I take to be the meaning of the *dark* colour of the dorsal and flank regions, as contrasted with the abdominal coloration, which I take to be a vestige of the ancestral *unarmoured* surface. The contrasted colours may be black, tan, brown or white (brown being a combination of tan and black).

That curious flank band which some mammals have, clearly seen in certain Squirrels, may possibly be a vestige of the receding carapace in process of total obliteration, and the different contrasted colours may indicate the different ages of partially unarmoured surfaces—that is, the abdominal colour would indicate the oldest unarmoured surface, the flank band a later unarmoured surface, and the dorsal colour the latest of all. The origin of this flank band is, however, not so clear to my mind as the other parts of this theory. I think we will find it impossible to account for *every* line, or band, or spot in the coloration of mammals, because the modifications in their coloration have been very great.

We cannot call the whole self-coloration as the youngest of all, for there may have been selfs among marsupials very much earlier than among other orders. Each group of mammals should, I think, be studied by itself, with an eye to its descent from some sort of carapaced ancestor.

It is a marvel to me that so much has been left to us of these hieroglyphics in which ancestral characters had been written, and that they have been so stamped in the nerve-centres that even time has not succeeded in wholly defacing them!

It is a marvel to me also that marsupials, being the oldest mammals, have *anything* left on their exterior to give any indication of their descent from, as I think, the common stock of *armoured* mammals!

Of this there cannot be much doubt. Armoured vertebrates—Fishes, Saurians, Chelonians, Glyptodonts, etc., formed, at one time of the earth's history, a vast portion of its inhabitants.

In the *Royal Natural History*, vol. iii. p. 64, indication is given of the existence in former times of some sort of armoured Dolphins. Under extinct Cetacean-like animals—*Zeuglodonts*, it says, 'So far as they can be determined, the general characters of these Zeuglodonts are such as we should expect to find in an ancestral group of Cetaceans; but it is remarkable that the body appears to have been protected by an armour of bony plates.' (!)

FURTHER EVIDENCE IN SUPPORT OF THE THEORY THAT EXISTING MAMMALS DESCENDED FROM CARAPACED ANCESTORS

'Nothing impresses the stamp of truth upon an hypothesis more than the fact that its light renders intelligible not only those facts for the explanation of which it has been framed, but also other and more distantly related groups of phenomena.'

Professor WEISMANN, *On Heredity*, p. 336.

PART IV

FURTHER EVIDENCE IN SUPPORT OF THE THEORY THAT EXISTING MAMMALS DESCENDED FROM CARAPACED ANCESTORS

THERE is yet another important feature in the coloration of animals which, if I have rightly interpreted it, will further support the theory I have been advocating in the foregoing pages—viz., that the markings of Leopards, Horses, Zebus, Giraffes, etc., etc., are due to the plate-armour of remote ancestors.

In these pages, however, I shall deal with mammals which may have no vestige of ancestral rosetting, but only a great *contrast* of surface coloration.

It has been maintained by some that the backs of certain animals are darker than their abdominal surfaces, because their backs are more exposed to light.

I do not know whether this is true of all fishes. Recent experiments would tend to show that light, as an electric agent, has something to do with the darkening of the skin of fishes.

I do know, however, that this is not true of many mammals. A few examples will clear this misconception.

The family of the Badgers and allied species affords a number of examples of black *under* parts, and either white or light-coloured *upper* parts; and most of these are *nocturnal* animals, so that we cannot consider their contrasts of coloration as having had much to do with a natural selection origin, by way of warning colours.

Then in the Natural History Museum we find that *Propithecus coquereli* has the front of all four legs brown, and the posterior of the legs and *back* of the body wholly white; that *P. coronatus* is brown on the abdomen and white on the back; that *P. Edwardsii* is dark brown everywhere except the posterior half of the back, which is white; that *Varesia varia* is black on the abdomen and front of legs, and yellow-brown on the back.

Further, the Egyptian Zorille or the parti-coloured Weasel has a white back, striped longitudinally with intense black, and its ventral surfaces and legs are black.

These might be sufficient to prove that light cannot be the cause of this difference in the colorations of the upper and lower surfaces of mammals. But if, in addition, we take that large class of mammals, which have a dark back and a lighter-coloured abdomen, we shall find that, in many cases,—such as in the Gazelles, the Black Buck of India, the black-backed Jackal of S. Africa, etc.,—there is a distinct and *very sharp* line of demarcation, along the flanks, between the two colours. This sharp line of demarcation alone would be wholly opposed to the theory under review, for there is no known property of light by which such a result could be produced on the flanks of these animals. Moreover, the front aspect of the legs of these animals is in most cases darker than the hind aspect; and we cannot suppose that the sun rays always beat upon the legs from the front. It is obvious that in walking away from the sun the *hind* aspect would be more lighted! The same objection can be brought against this view, in the case of those Antelopes which have patches of white either in front or behind, and which Dr. Wallace would call 'recognition marks.' Then what shall we say of all those Antelopes and Deer which, on a dark ground, are either spotted or striped with white? Why have these white marks not been obliterated by the light?

MAMMALS WITH CONTRASTED COLOURS 139

What I have said is enough, I think, to prove that this light theory is wholly untenable, at least as far as mammals are concerned.[1]

The reader might however say, Have you any other theory to substitute for the light theory?

Yes, I have, and it is this: I have endeavoured to show that the spots and stripes of mammals are in many cases *pictures* of ancestral *armour-plates*, modified subsequently in hundreds of ways in some, or wholly obliterated in others.

Well, the same theory of descent from armour-plated ancestry will explain why there is on the flank of many animals that sharp line of demarcation between the dorsal and the ventral colouring. That sharp longitudinal line indicates the *margin* of the ancestral *carapace*, as we see it in the Armadillo, the Glyptodon, the Pangolin, and others.

Curiously enough, the little Armadillo of Argentina (*Chlamydophorus truncatus*) has its armadilloid armour sharply truncated posteriorly, and its rump is patched up by a differently plated perpendicular shield. This curious appendage, according to my view, might perhaps account for that white patch which so many ruminants have behind, the line of demarcation between the dark

---

[1] I am aware that in the cases of peaches, apples, pears, etc., light colours their cheeks, and I have seen a peach, partly shaded by one leaf, not acquire colour under the leaf. But I am also aware that round the stone of certain peaches there is the same red colour we see on their cheeks, while the intermediate thickness (sarcocarp) is either white or yellow, and this central redness could not be produced by light in its ordinary sense. Then there are the purple aubergines, which acquire their intense colour also on the shaded sides. And I have been informed by a noted horticulturist that he has seen the black Hamboro' grape beautifully coloured when the roof of the house was almost wholly shaded by the leaves. He also told me that he has seen the red seakale and the red rhubarb grown in total darkness, and the tips of their leaves nevertheless always acquire a red tinge. Moreover, we do not see the sweet-water grape colour, although grown outside and fully exposed to light. I think if we substitute electricity for light, acting on certain chemical ingredients, we may perhaps be nearer the right explanation.

back and light-coloured rump being very sharp. It might mean that the ancestors of those ruminants *had* that rump shield, and that it was got rid of *earlier* than the dorsal carapace, and thus a sharp contrast of coloration was inherited in consequence. There is an alternative supposition, for in some Armadillos we find that the carapace stops short of the tail, and leaves a bare space between the tail and posterior margin of the carapace, and this fact may remain recorded in the ruminants and others that have a white patch behind. It would seem to mean only a continuation of the unarmoured ventral surface.

Among the animals that are so patched behind are the Siberian Roebuck (*Capreolus pygargus*), Wapiti Deer (*Cervus Canadensis*), Pronghorn (*Antilocapra Americana*), Bonte-bok (*Damalis pygarga*), Sable Antelope (*Hippotragus niger*), Sœmmering's Antelope, and several others. It may be quite true that animals now make use of these white patches as 'danger marks,' and 'recognition marks,' but their origin is I think explained by the abruptly truncated dorsal armour of some ancestral form. In the 'Pichiciago' Armadillo of Argentina, the vacant space is covered by an additional rump-shield, but in others it might be absent, as we indeed see it in several other varieties of Armadillo.

In this connection I would mention that in the Natural History Museum there is an interesting and very suggestive preparation of this little Armadillo. It has no calcareous or bony carapace— but only the epidermic horny shell of one. Under this horny shell there is a hairy coat as in other animals. The curious part is, that the horny shell is attached to the true skin by means of a thin ligament all along the spine. This latter feature may perhaps give us a clew to the interpretation of that spinal dark or white[1]

---

[1] The Kerry breed of cattle are black, with a white streak down the back, and sometimes another along the belly.—(*Roy. Nat. Hist.*, vol. ii. p. 168.)

stripe which so many animals present. This may perhaps be a *vestige* of the attachment of this dorsal ligament in some ancestral form, during the *transition* between the stage of plate-armour and the stage of hair-covering. The spinal dark or white stripe of certain animals might indicate that the general colour of the surface altered *earlier* than the spinal surface, and therefore had time to become contrasted in colour. I will not however press this point, should the reader think it preposterous. It may be looked upon as a mere suggestion.

My contention is that the differently coloured dorsal and flank regions of certain animals, as contrasted with the coloration of the abdominal surface, is a *vestige* of the ancestral *armoured* carapace, the sharp line on the flank indicating its margin, while the white, or differently coloured abdomen, as in the Badger, the Egyptian Zorille, and others, would indicate the *un*armoured ventral surface of the ancestral forms. The origin of the white patch behind would fit into this theory, and would indicate the truncated posterior margin of some ancestral carapace.

In studying the Armadillo, we find that the abdomen, legs, and under surface of the head and the throat are almost denuded of scales, and much more so are the hind-legs of Kappler's Armadillo.

Then in some Pangolins, we also find that the under surface of the head, the throat, the inner aspect of the legs, and the abdomen, are *devoid of scales*.

It can readily be understood that scales on those parts would prevent the animal from rolling himself up for defence. The loss of scales in those parts, while it gave him greater *freedom of movement*, enabled him to roll himself up, and *protect* those unarmoured surfaces, just as a Hedgehog would do.

Now in the Black Buck (*Antilope cervicapra*), instead of armoured and unarmoured surfaces, we find simply contrasts of coloration in homologous surfaces, viz., the forehead and nose, the body and outer aspect of the legs, and upper surface of the tail are black, and the under surfaces, corresponding with the unarmoured surfaces of the Pangolin, are *white*. This, in my judgment, means that the ancestor of the Black Buck lost its armour on what are now white surfaces *long before* it lost it on what are now dark surfaces, so that those had the opportunity of having their pigment modified,[1] and so becoming contrasted by the time the descendants wholly lost their armour. And the only thing now left to tell the tale of ancestral armour in many animals is solely this *contrast of coloration*, which in certain animals still remains, as a *survival* of a feature which in remote times may probably have been *general*.

In the Horse, through innumerable selections, this contrast of upper and lower surfaces has been mostly extinguished, although Horses of a brownish colour are often met with which have the lower surface much lighter than the upper. But it is in the costermonger's Donkey that a great contrast is visible between the dark back and flanks and white abdomen. In black-and-tan Dogs the contrast between the two surfaces is very marked; and there are variations in which the contrasted colours of the Dog are *tan and white*, or black and white.

It is not improbable that as time goes on all traces of contrast derived from armoured and unarmoured surfaces may disappear, unless maintained for special purposes, either by natural or artificial selection, or unless reversions to ancestral colourings occur. Among domestic animals this obliteration has already largely occurred. In wild animals we see this obliteration in the Lion, the Puma, the Hare, certain Deer,[2] and others.

[1] In the domestic Dog we see how readily the hair pigment is modified, even in one generation, from black to tan, and from black or tan to white.

[2] The young of the Wapiti Deer is spotted, while the adult has no trace of spots. —*Cervus Canadensis*, Science and Art Museum, Edinburgh.

Some Horses, as I said, show a light-coloured abdomen and inner aspect of legs, but this is not common. Usually the lighter colour is very partial, and nothing like what it is in the domestic Ass, or the Black Buck. See Grevy's Zebra, fig. 52.

Dr. Wallace, I think, is partly right and partly wrong, when he says that the Rabbit has acquired the white colour on the underside of its tail by natural selection, so that it might use it as a danger-signal. In my judgment it got the white colour by *inheritance* from a remote ancestral *unarmoured* surface, the white colour being a *vestige* of that ; while it may have acquired the habit of turning up its tail, and showing the white banner, through *natural selection*. We know, too, that other animals, under the excitement of fear or anger, cock up their tail, so that it is no wonder the Rabbit should do the same. The survival of those that could perform this social function would follow as a consequence. The Rabbit is a defenceless and timid animal. For its safety it has to depend on its large ears, in addition to the faculty of running into holes when its ears declare that possible enemies are about ; and turning up its white surfaced tail, when running home in the uncertain light of the dusk, may be a very useful way of showing its associates which way the holes lie.

This theory I am advocating will also explain why the front or exterior aspects of all the legs of many animals are so often, especially in Gazelles, differently coloured from the posterior or inner aspects. Naturally the front and outer aspects of the limbs, which received the brunt of the enemy's attacks, would have been *differently* armoured from the posterior and inner aspects. This is the state of things in certain Pangolins, for instance. Something may perhaps also be due to the likelihood of there being only a limited amount of armour-material in the blood, and only those parts which urgently needed protection could be supplied with it.

The reader should understand that, as in the case of spotting, innumerable modifications have occurred, through which, in many cases, the sharp line of demarcation, between the ancestral armoured and unarmoured surfaces, has been toned down into a graduated passage from the dark dorsal to the light ventral colouring.

This theory, however, should not be driven too far, for it will not account for *every single hair* that may be differently pigmented ; but I think it will account for the general colouring and spotting more satisfactorily than the older light theory. Then human selection of animals under domestication, and natural selection of animals in a wild state, will account for the vast number of modifications we see in the coloration of animals.

I would ask the reader to glance at Fig. 72. Nothing, to my mind, can be plainer than the tale its coloration and spotting suggest regarding its descent, in spite of the great modification its legs and its digestive apparatus have undergone to fit it for surroundings totally different from those of its ancestors. Its partially edentate upper jaw seems to tell a similar story. This Deer would seem to say, ' My spots are the vestiges of ancestral armour-plating, and that fringe you see on my flank is the vestige of the fringe of differently-shaped plates which form the margin of the Glyptodon's carapace.'

No one could look at the picture of *Mellivora Indica* in Mr. Blanford's *Mammalia of India* (Fig. 46), without seeing in its well-defined grey back a strong resemblance to a carapace ; and the picture-carapace of *Putorius sarmaticus* (Fig. 41), is still more striking, as its vestigial carapace still retains the vestigial spotting of its ancestral bone-rosettes.

Even the sheep in the London parks seem to tell you, ' My fleece is the substitute of my ancestor's carapace. In those days, my ancestor's face, ears, hands, and feet, were unarmoured, and now

MAMMALS WITH CONTRASTED COLOURS   145

the hair *there* is short, and of a different colour from my fleece—that is either white, brown, or black. My relative the Wolf, who in former days was my enemy, is now my friend and caretaker. He

FIG. 72.—A Spotted Deer, from a photograph, C. R. 1393.

too in his "fleece" shows distinct evidence of our common ancestor's carapace.'

If we turn to the White-backed Skunk, a good picture of which is given in the *Royal Natural History*, vol. ii. p. 76, we find

K

similar contrasts. The peculiarity of its contrasted coloration, the writer says, is regarded as belonging to the class of so-called 'warning colours.' Mr. Poulton observes that such warning colours would seem 'to benefit the would-be enemies rather than the conspicuous forms themselves. . . . . But the conspicuous animal is greatly benefited by its warning colours. If it resembled its surroundings, like the members of the other class, it would be liable to a great deal of accidental or experimental tasting, and there would be nothing about it to impress the memory of an enemy, and thus to prevent the continual destruction of individuals. The object of warning colours is to assist the education of enemies, enabling them to easily learn and remember the animals which are to be avoided.'

There cannot be much doubt that an animal possessed of such a coloration and character, and also of such 'nauseous and irritating artillery,' as has been described, *would* be avoided, when its means of defence had become known. But we cannot in any way admit that it is its 'stinking secretion' which has *caused* this 'contrast of coloration.' The animal itself no doubt has learnt that it is not attacked, and this accounts for its 'indifference to the presence of other creatures' which is said to be 'one of the most striking characteristics of this animal and its congeners.' Therefore the contrast of coloration between the white upper surface and crown, and the black under surface and face, must be attributable to some *other* cause.

When the plan of coloration of this *White-backed Skunk* became once established by *hereditary influences*, it began to change like everything else, *unless* it were *maintained* by natural selection, through the action of the surroundings, as has happened in the case of the Leopards; and this change we see, further on, in a brother Skunk.

We can understand that an animal something like a Pangolin, when it got rid of its armour, and when hair was substituted for it,—as indeed we see has occurred in the little Pichiciago—the hair covering remained under the same nervous influences which by habit its ancestors possessed when they *had* back-armour ; and in all probability the contrast of coloration of the hair of some of the descendants is caused by the same nervous influence.

As a matter of fact we find that another animal, which we might say is a *brother* of the Skunk we have been discussing, has not this conspicuous warning coloration, *although it possesses a similar* ' nauseous artillery.' This *Common Skunk* (so different from the *White-backed Skunk*) has a black or blackish body, and 'although there is a great amount of individual variation, the white markings usually take the form of a streak on the forehead, a spot on the neck, and two stripes running down the back. . . . In some cases the white stripes do not extend beyond the neck, so that the back is entirely black.'

In the Common Skunk, it would seem, the change of colour has gone on to such an extent as to leave nothing but mere *vestiges* of its ancestral carapace-like *white back* ; and therefore the theory of warning colours having been caused by ' mephitic ' influences evaporates.

It can be readily understood that possible enemies who may come within the sphere of its effluvium don't require to *see* any warning colours, they can *smell* the animal from a long distance, and would naturally leave a good space between him and them ; so that the ' mephitic ' warning may continue without the so-called *warning colours*.

One can hardly contemplate the Black-backed Jackal of South Africa in the Zoological Gardens and not think its sharply defined back related to an ancestral carapace. The Dingo of Australia has similar markings, but the black on its back is less marked. Very probably, either the one or the other has given our black and tan domestic Dogs their distinctive coloration, which, in some varieties, becomes *tan and white*. These and many others, in my opinion, owe that particular feature to an ancestral carapace, although the separate spots—vestiges of bone-rosettes—may have wholly disappeared; they still persist, however, in the Dalmatian breed of Dogs.

I have seen a curiously marked Toy Terrier of the black-and-tan breed. Its back was grey and sharply defined like that of a Badger, and it was blotched and striped with black. And in the Natural History Museum there is a tiny Cheetah from the Cape which is also curiously marked. Its back is grey like that of a Badger, and the other parts are spotted. It is impossible to contemplate these reversions without thinking that they must have a deeper meaning than that of being simply accidental.

At the risk of wearying the reader with repetitions, let us now try to recapitulate briefly the whole process, and endeavour to form a clear conception of how these phenomena could have been brought about.

First, we must assume that natural selection, as indeed is admitted by modern biologists, must have had a great deal to do with evolving, from previous modes of armour, with the help of congenital variations, the forms of carapaces we see in the Glyptodonts and Armadillos with plate-rosettes. The congenital variations in the plates of carapaces were presumably brought about by changes in the central nerve-action. This action we as yet do not understand, any more than we understand how thought is evolved from it. There can be no doubt however that in the higher animals it is all-important, and governs everything.

This original nerve-action, however caused, by which a rosetted

MAMMALS WITH CONTRASTED COLOURS 149

carapace was elaborated, became confirmed by ages of usefulness. In other words, the nerve-centres acquired the *habit* of acting that way.

Then when, from whatever cause, the calcareous matter of the exoskeleton disappeared, the nerve-centres *continued* that same action, which resulted in *pigment pictures* of the ancestral plate-rosettes on the supple and elastic unarmoured skin. This is not all, for the *margin* of the carapace was also pictured by contrast of colour between the upper (ancestrally armoured) and the lower (unarmoured) surfaces; the margin *remaining* pictured even when *all traces of rosetting had disappeared.*

Then innumerable further modifications in the atomic constitution of the nerve-centres, in which natural selection no doubt has played a great part, have resulted in all the varied colorations of mammals we see.

Of course it is impossible to say what atomic changes in the nerve-centres produce a dark back and a light abdomen in some, or a light back and dark abdomen in others, any more than it is possible to make out why one animal turns out albino, and another melanoid, or why in one very dark grey Horse, with only few vestiges of spots, its mane and tail were *pure white.*

Some might perhaps fancy that they account for changes in coloration by saying, 'Oh! that is an *albino*,' or 'that is a case of *melanoid* variation.' But this in no way explains its cause any more than if one said it is a 'whitino,' or a 'blackino'! We are at present wholly unable to say why, in a litter of black-and-tan puppies, one comes out wholly tan, or tan and white; or why, in a litter of tan puppies, one comes out wholly black, or wholly white. All we can say is that these phenomena *do* occur.

This we may say; in the Jaguar these changes of colour occur in the components of the *rosettes themselves*. What I conjecture was ancestrally a big central plate is now in the rosette of the Jaguar a *brown patch*; and what were the encircling platelets are now *black spots*, or a fusion of them forming a black ring, the whole rosette being of a different colour from the general fawn colour of the inter-rosette spaces. These three distinct colorations, well marked in the Ocelots, suggest atomic *localised* differences in the nerve-centres of which at present we know nothing.

Why in the Cats the spots are black, and in the Deer they are white, I am unable to say. I find it also impossible to determine whether the grey dappled Horse is an *albinoid* variation of the brown dappled Horse, or the latter is a *melanoid* variation of the former. I must leave such questions for others to answer who may know more about the matter than I do.

All I am here concerned with is, that there *is* in many animals a sharp line of demarcation between the colouring of the back and that of the abdomen, and this I attribute to *ancestral differences of armoured and unarmoured surfaces.*

No one who is imbued with the principle of evolution can contemplate the skin of the Jaguar in Fig. 4 without saying, Yes; the whole thing is a picture of armour-plating. This is not, however, all, for that skin, in the different mode of rosetting between the dorsal and the ventral surfaces, and in the difference of *general* coloration between the upper and lower parts, gives also evidence, at least to my mind, that the lower lost their armour *before* the upper parts.

This theory does not attempt to account for the variations of nerve-centres. Those must be relegated to congenital causes we do not yet sufficiently understand, and perhaps to the action of the environment.

In another part I shall endeavour to show that armoured mammals must have been originally and ancestrally armoured all over, that is, dorsally and ventrally, like the extinct Ganoids and existing Crocodiles ; and that in dispensing with their armour they *first* lost it on their ventral aspect, and *second*, and much later, on their dorsal aspect also.

At the risk of seeming tedious, I repeat that this, in my opinion, is the reason why so many mammals present a distinct contrast between the coloration of their dorsal and ventral aspects with a *sharp* line of demarcation between the two. Of course later on there may have been a *fusion* of the two distinct colorations, and all traces of demarcation wholly obliterated. Examples of this obliteration are seen every day among Horses, Dogs, etc.

RESEARCHES AND DISCUSSIONS TO
CONNECT, MORE SURELY, ARMOUR-
PLATING WITH SKIN-PICTURING

'We can only say generally, with Darwin, that selection works by the accumulation of very slight variations, and conclude from this that these "*slight variations*" *must possess selection-value*. To determine accurately the degree of this selection-value in individual cases is, however, as yet impossible.'

The All-sufficiency of Natural Selection, by Professor AUG. WEISMANN, *Contemporary Review*, September 1893, p. 322.

PART V

RESEARCHES AND DISCUSSIONS TO CONNECT, MORE SURELY, ARMOUR-PLATING WITH SKIN-PICTURING

Is there any tangible evidence to prove that skin-spotting is often the result of ancestral armour-plating?

To answer this question we have to take a wider view of vertebrate animals.

There is a great number of existing animals, which now have only a partial and scattered armour, evidently a mere vestige of a more complete and closer fitting ancestral armour.

The partial or complete disappearance of the bony plates of the exoskeleton, from whatever cause, not unfrequently leaves *spots* or other marks on the skin, in their stead, as records, so to speak, of what had gone before.

I shall give only a few instances from fishes, which can be seen in the Natural History Museum and in other records.

A Ray-fish with a whip tail, labelled *Urogymnus asperrimus*, has a broad carapace, from head to tail, studded with closely-set plates of two sizes, viz., large stellar and spinose plates, encircled by minute tubercular platelets, the tail being wholly encased in similar plates.

Then an allied Ray-fish (a sting ray) from the Australian waters, *Trygon tuberculata*, has only scattered spinose plates on its head and sides, and a complete spinal line of similar plates, running into the tail, which is also covered with spinose plates.

Of another allied species, also from the Australian waters,

Trygon brevicaudata, there are two specimens, (*a*) with a smooth skin, seven plates on dorsum of tail, and minute scattered plates on other parts of the tail; (*b*) also with a smooth skin, but with several *spots* on both sides, and only eleven plates on dorsum of tail besides other minute tail-plates.

This series of fishes is very instructive, and there is no good reason to doubt that in the latter the spots are partial vestiges of the more complete armour of *Urogymnus asperrimus*. The back-armour and tail-armour of the latter would lead us on, in the course of ages, to the huge carapace and curious tail-armour of the Glyptodonts.

There are many spotted Sharks, and the spinous Shark (*Echinorhinus spinosus Gmelin*)[1] is dotted with spinous plates, some isolated, others confluent in groups. It seems probable that these are only *vestiges* of an ancestral and more complete carapace.

Then there is another fish in the Natural History Museum which, in this connection, is very instructive. It is the *Serranus gigas* of Muscat. It is a large scaly fish in one of the middle cases. The horny scales of fishes can only be considered as a modification of the bony armour-plates, like those of the bony Pike and extinct Ganoids. Well, the large body scales of this *Serranus* have each a *black dab*. On the cheeks of the fish[2] the scales become thinner and thinner, and almost *amalgamated* with the skin, each scale still retaining the black dab. Then on the lips of the fish the scales disappear, but the black dabs or spots *remain*.

Of course the spots in this case are not exactly identical in *genesis* with those of the Jaguar skin, unless we suppose that the Glyptodontoid ancestors of the latter, besides their bony plates, had also a corresponding set of horny plates, like those of Armadillos, and spotted analogously to the spotting and marking of the

[1] Science and Art Museum, Edinburgh.   [2] Technical terms are here unimportant.

ARMOUR-PLATING AND SKIN PICTURES 157

carapace of certain land Tortoises. The black dabs on the scales of *Serranus* are as distinct from the skin as the ocelli in the Peacock's feathers, yet where the scales are suppressed, the *black dabs remain*.

A striking example of stamps remaining on the skin after plate-armour has disappeared is to be seen in *Serranus hexagonatus*.[1] All over the surface it has hexagonal markings independent of the scales. These markings forcibly recall the hexagonal armour-plates of the Ostracions. Then *Serranus Merra* has the marks passing into confluent blotches, while *S. Sonneratii*, Pl. vii., Fig. 1, has similar hexagonal marks, but only on the fore part of the body and head, being totally obliterated on the rest of the body.

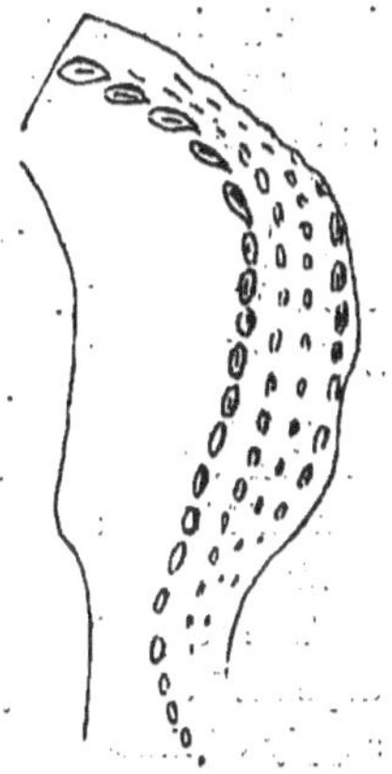

The singular markings of these species of Serranus are quite paralleled by the markings of the fully dappled and the partially dappled Horses.

FIG. 73.—Back of *Sclidosaurus Harrisoni*, as restored in Hutchinson's *Extinct Monsters*, pl. 8.

Then in the York Museum I found an Ostracion, with plates shown in Fig. 66 (*b*). They were marked with six dark spots. Well, this fish on its tail had no plates, *but only the black spots*.

Then if we look at the back of the Dinosaur, with rows of detached plates along its spine, as restored by Mr. Hutchinson (Fig. 73), we shall see that if the plates disappeared, rows of spots might be the result, like those along the back of certain Leopards.

I know that there are puzzlers among the markings of fishes, such as the Leopard markings of *Trygon uarnak*, and the Giraffe

[1] *Fishes of India*, by Francis Day, Pl. ii., Part. 1. Fig. 3.

markings of *Murœna tessellata*, and others, but one cannot be expected to tackle the whole of creation at once, and the interesting spottings and markings of fishes must be left for some future investigation. By that time experimenters on the spotting of lower animals may perhaps be able to give us something definite as to their cause.

This I know, that, even in plants, hairs or spines are often accompanied at their base with spots, and when the hairs disappear the spots remain. Here are a few instances: *Begonia argenteo-guttata* has spotted leaves, and each spot has a hair, which is the plant armour, while in *Begonia Rex* the leaves are piebald, that is, a number of spots have become confluent. Some varieties of this remain haired all over, while others have lost their hairs. Then in *Begonia maculata* the spots remain, but the hairs are suppressed.

The deposition of dermal plates in animals undoubtedly depended on heredity, the action of nerve-centres, and an available supply of lime-salts, so that, as I have stated further on, when the plates ceased to be developed, from a deficiency of lime-salts in the blood, the nerve-centres, still continuing their influence, would have brought about changes in the pigmentation, which now, in certain cases, *picture* the plates themselves.

The main evidence, however, of the Jaguar and Leopard having descended from plate-armoured ancestors is in the resemblance of their pigment-rosettes to the plate-rosettes of those extinct mammals;[1] the same nerve-action which gave rise to bone-rosettes was certainly equal to produce pigment-rosettes, from which the hairs grow. That hairs will often grow as soon as there is a part of the skin free from plates is sufficiently proved by the Hairy Armadillo, and some of its congeners.

[1] In the Tring Museum, as already mentioned, there is a Jaguar which, on the flanks, has *very* large polygonal rosettes with *many* (from one to six) spots in the enclosed space.

ARMOUR-PLATING AND SKIN PICTURES

Indeed, Professor Parker[1] has shown that in the embryo of the Pangolin the scales are nothing but matted hairs, and the intervening spaces are also covered with hairs, so that the substitute for bone and horn armour, in mammals, seems to be hair or wool.

The description of spotting phenomena in animals by words is a feeble thing compared with illustrations by photography, but to photograph every modification would require a library, and not one book.

The *general* colouring of a mammal is of little importance, because we know that this varies very much; but if the markings are constant, though the general colour changes, we must infer that they have a different and a deeper meaning than the general colouring.

In Pigeons, the wing-bars very often remain, although the general colouring may vary *ad infinitum*. There are grey and also cream-coloured Pigeons with brown bars, and blue Pigeons with black bars, and so forth. And Dr. Wallace mentions[2] that in the skirts of the forests on the Amazon, and in the larger 'ilhas,' both the black and spotted Jaguars are often found. By black, I presume he means the melanoid variety with distinct rosettes, which can readily be distinguished in certain lights. We see a similar persistence of markings in the Snow Leopard, although the ground colour has changed to white.[3]

From all this persistence of markings, as something distinct from the general colour, I would infer that the markings have a *deeper* meaning than the general colour, and the meaning I would attribute to these persistent markings in mammals is that they have been inherited from *much more remote ancestors* than those

---
[1] *Mammalian Descent.*   [2] *Travels on the Amazon*, p. 63.
[3] In the Tring Museum I saw two varieties of Snow Leopards; (*a*) with *ocellated* rosettes, and (*b*) with a large number of the rosettes *solid*, especially on the shoulders, haunches, and lower flanks.

which have given them their general colour, that is, from an ancestry in which plate-armour had been *established for ages*.

I would repeat that the *spotting* is the important feature, and not the *colour* of the spotting—for we see in the Cheetah, black spots on a light ground, and in the Deer, white spots on a dark ground, and in the Kerry Cattle we see a *white* spinal line, instead of the ordinary dark one. The reader might say—Is it reasonable to suppose that an armour-plate should leave an impression after the bony plate had totally disappeared, and should continue to be pictured in the descendants for innumerable generations? It does seem strange that this should be so, but impressions are left from far more transient causes than the carrying of bony plates on one's skin for perhaps millions of years.

Mr. W. B. Croft, in a communication to the Physical Society,[1] has shown that impressions of coins can somehow be left on a *clean glass*, invisible at first, but made visible by breathing on the glass; also that an impression of a paper, printed only on one side, can be invisibly taken by pressing it between two plates of glass. The printing can be brought out by breathing on the glass which was opposed to the printing. What is more curious is that a similar print-impression is given to the glass which does *not face* the printed surface. This latter impression can also be brought out by breathing over the glass surface, and it is evidently produced *through* the paper. Presumably light may have something to do with these impressions. But if so slight and temporary an influence can leave an impression on the glass, what wonder would it be if the armour-plating, carried for who knows how many ages, should have so modified the skin, and its nerve-centres, as to transmit plate-*pictures*, even when the calcareous plates had totally disappeared?

[1] Reproduced in *Year Book of Science* for 1892, p. 16.

ARMOUR-PLATING AND SKIN PICTURES 161

My object in all this discussion has been to endeavour to discover why the Tiger and Zebra are striped, why the Jaguar and the Leopard are rosetted, and the Deer and Horse dappled, and why there are so many animals of different classes which have ringed tails. The skeletons of all these animals inform us that they are *all* related structurally, and therefore they must have come from some common ancestral source, if not from the same pair of parents, certainly from the same class of parents.

In another place I have shown that the same pattern of armour-plates can be traced, through the Crocodile to the Sturgeon, and perhaps also much lower down in the scale of life.

Professor Huxley[1] says: ' If the doctrine of evolution be true, it follows that, however diverse the different groups of animals and of plants may be, they must all, at one time or other, have been connected by gradational forms ; so that, from the highest animals, whatever they may be, down to the lowest speck of protoplasmic matter in which life can be manifested, a series of gradations leading from one end of the series to the other, either exists, or has existed.'

Professor Alleyne Nicholson in his *Swiney Lectures* of 1893 repeated the same thing. He told us that all forms of life on this earth originated from pelagic low forms like those still in existence in the oceans.

But what is there to show that in the picture-plating of the Jaguar the enclosed space is *homologous* with the larger middle bone-plates of the Sturgeon, the Crocodile, the Glyptodon, etc. ? There is nothing to show this beyond the difference *in colour of the enclosed space* in the Jaguar rosettes as compared with the *general* ground colour of the animal's skin. I have endeavoured to emphasise this difference of colour in No. 9, Fig. 59. The shaded space

[1] ' Science and Hebrew Tradition '—*Lectures on Evolution*, p. 89.

L

between the rosette blotches indicates the enclosed space, but the Jaguar skin of Fig. 4 sufficiently shows that the enclosed spaces of the rosettes are of a different and deeper shade.[1] In some Ocelots the enclosed space is quite brown. This difference of colour would certainly indicate that in the nerve-centres there was some difference, although of an atomic .character, in the cells which regulate the colour of the inside and of the outside of the rosettes. In all probability the atomic difference is of the same nature as that which caused a large plate to be deposited in the centre of the armour-rosette and a ring of small ones outside it. What we have to note is that there *is* a difference between the inside and outside of the Jaguar and Ocelot rosettes.

Why this is so we do not know, any more than we know why one Horse is dun, another bay, a third brown or black ; why a dun, a bay, a dark grey, and even a black Horse sometimes has a pure white mane and tail, and so forth.

To sum up, in the existing Jaguar coloration we have the following elements, no doubt much modified from those of the ancestral Jaguar :—

(*a*) We have the general ground colour of a rich tan between the rosettes ;

(*b*) We have the spots more or less fused, which make up the polygonal rings of the larger rosettes ;

(*c*) We have the enclosed space, which is often of a *darker* hue than that of the ground spaces ;

(*d*) We have those curious central black spots.

Now I have elsewhere mentioned that tan colour is interchangeable with either white or black, and so we have the general colour changed into either brown or black in the black Jaguar.

---

[1] In the photograph this is decidedly shown, but in the illustration the difference of colour is not so clear.

ARMOUR-PLATING AND SKIN PICTURES 163

The enclosed space *alone* may change into black, and so obliterate the space and turn the ocellated rosette into a *solid* rosette, as we see it in the Serval (Fig. 17); or the rosette may be so contracted as to obliterate only the middle spots, as in most Leopards. Finally, we may have the *whole* surface changed either into a jet black, obliterating all traces of rosettes, as in the black Leopard of Johore, and the jet black domestic Cat; or the surface may be changed into a uniform rich isabelline colour, as in the adult Puma; or the ground colour inside and outside the rosettes may change to white, as in the Snow Leopard. A further obliteration of all colour produces the albino domestic Cat.

In all this theorising about the descent of rosetted, spotted, and striped animals from carapaced ancestors, it should be distinctly understood that I do not in the least maintain that the Jaguar or Leopard descended from *this* particular *Glyptodon*, or *that* particular *Dædicurus*. What I mean is that the Jaguar and Leopard bear on their skins the *stamp* of having descended from *a* carapaced ancestor, which had bony rosettes something like those of a Glyptodont. How many bone-rosettes, and of what exact shape they were, in this conjectural ancestor, I am not in a position to say.

In this discussion the whole evidence is circumstantial, for no one has ever seen the passage of a Glyptodon's carapace into the rosettes of the Jaguar. One would indeed require to have lived a good bit of time to witness a Glyptodon changing into a Jaguar, considering that Cuvier found no appreciable difference between the skeletons of the ancient mummified animals of Egypt and their representatives which lived 3000 or 4000 years later.[1]

In these and similar investigations circumstantial evidence is the only kind of evidence obtainable, and it is very valuable. The connecting gradations must be filled up by the imagination.

[1] 'Science and Hebrew Tradition'—*Lectures on Evolution*, p. 77.

PROBABLE MEANING OF SOME INTEREST-
ING FEATURES IN HORSES AND
OTHER MAMMALS

' Positively, the principle may be expressed, in matters of the intellect, Follow your reason, as far as it will take you, without regard to any other consideration. And negatively, in matters of the intellect, Do not pretend that conclusions are certain, which are not demonstrated, or demonstrable.'

From Professor HUXLEY's *Definition of Agnosticism.*

PART VI

PROBABLE MEANING OF SOME INTERESTING FEATURES IN HORSES AND OTHER MAMMALS

BESIDES the dappling which I have tried to account for, Horses have certain other marks which are very common indeed, and which are more conspicuous in self-coloured Horses, such as bays, browns, blacks, etc.

For instance, Horses very commonly have what is called a white 'blaze' all over the front of the face down to the mouth; or a simple white star on the forehead, which varies between a mere white speck and a white patch.

Then there are intermediate stages as shown in Fig. 74. Sometimes the 'blaze' is interrupted, a white patch being on the forehead, and another on the nose, and so on. Finally every trace of the white blaze may be obliterated, and the whole face is of the same colour as the body. Frequently, however, in dark and dun-coloured horses the blaze is black, and in dark grey horses it is of a sooty grey, more or less interrupted, on a lighter ground.

Now, is there any way of accounting for this feature, so common in the domestic Horse? One day, in Piccadilly, I saw standing a dark grey cab Horse. On its forehead it had three pairs of faint radiating stripes of a grey colour on a white ground—as shown in Fig. 75 (*a*). It struck me that these might be vestiges of the Zebra face-marks. There happened at the time to be a stuffed specimen

168 STUDIES IN THE EVOLUTION OF ANIMALS

of Burchell's Zebra in Rowland Ward's window; so I went over to look at it, and found that the marks on this Horse's face corresponded to three pairs of similarly radiating stripes on the Zebra face, as shown in Fig. 75 (*b*).

In studying this curious feature further, I came to the conclusion that the dark blaze on the dark grey Horse, and on some bay and dun-coloured Horses, was a more or less complete *fusion* of the *stripes* we see on the Zebra's face, thus forming a dark blaze.

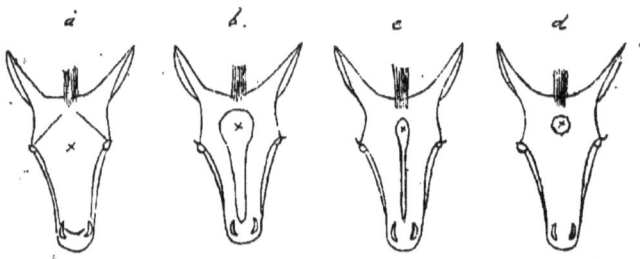

FIG. 74.—Diagrammatic sketches of the faces of Horses; (*a*) shows a full white 'blaze'; (*d*) a white star on the forehead; (*b* and *c*) intermediate contractions of the 'blaze'; (×) indicates the white part on a dark ground.

Then, as black is interchangeable with white, in some dark Horses we see a complete *white* blaze, occupying the whole of the lozenge-shaped face of the Horse, as shown in Fig. 74 (*a*). The other figures (*b* and *c*) are mere contractions of the blaze, until we come to a mere *vestige* of the blaze in the shape of a small white star between the eyes, which, in others, becomes completely obliterated.

It would be tedious to go through all the variations that this blaze is subject to; suffice it to say, that it sometimes invades the whole head, while in others it disappears completely.

INTERESTING FEATURES IN HORSES, ETC. 169

I have seen a Horse with a blackish star on a white blaze, and others, mostly bays, with a white star on a black blaze. This shows that although usually the star is white, it may, under certain circumstances, be exchanged for a black one; and although usually the blaze is white, it sometimes can also be black. I have seen a perfect *black* blaze on the face of a light dun Horse. It was as full as the complete white blaze of (*a*) Fig. 74.

These features in the domestic Horse are so persistent, even when all other spotting from the body has wholly disappeared, that

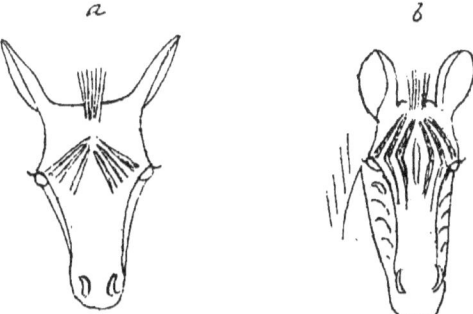

FIG. 75.—(*a*) Faint radiating stripes on the forehead of a dark grey Horse ; (*b*) black radiating stripes on the forehead of Burchell's Zebra—note its *lozenge-shaped* face.

they would seem to have some deeper meaning than mere *accidental* marks. It is interesting to note that even the Zebra may have a small black star on its forehead between the eyes, as seen in Fig. 50.

The persistence of these features may be because domestic Horses have been bred originally from parents that had them. Breeders of first-class Horses, I understand, would rather not have either stars or blazes, and it may be presumed that they have endeavoured to select them out. Yet it is only exceptionally that they have succeeded in getting rid of them, for the large majority of Horses of all kinds one sees in the streets of London have these frontal marks, either fully developed or with vestiges of them ; so that I think we would be right in concluding that they come from a *remote* ancestry.

'Our grandfathers have told us how their fathers expatiated on the merits of the Dutch Horses (old Lincolnshire blacks), of their size and feats of strength, how the blacks with white legs and blazes were most esteemed. These animals, or their descendants, in time became located all over England.'[1]

This would appear a sufficient reason why blazed horses are so common in the streets of London. They got it from the old Dutch Horses, 'which with white legs and blazes were *most esteemed.*'

But this is not all, for Mr. F. Finn tells us[2] that 'The Onager was put to an Abyssinian wild Ass, and produced a hybrid ; it then bore, to a male Onager, a chestnut foal with a white blaze on the forehead ; but as this foal thus resembled neither parent, and in fact exhibited a Horse's rather than an Ass's marking, the case is surely one of analogous variation.'

The names of the equines appear to me to be mere *verbal* distinctions used by systematists in classifying animals. In nature there seems to be no such distinction between Horse, Ass, Onager, Zebra, and Quagga, more especially as they all *inter-breed*. And this blaze on the forehead of the chestnut Onager's foal seems a reversion to some ancestral mark—white or black—whether we call the ancestor a Horse, an Ass, an Onager, or a Zebra. In this foal the blaze was *white*, but I have shown that the blaze may have originally been *black*, and caused by a *fusion of the stripes on the*

'Pedigrees of British and American Horses,' by J. I. Lupton, *Nineteenth Century*, June 1894, p. 926.

[2] 'Some facts of Telegony.'—*Natural Science*, December 1893, p. 437.

INTERESTING FEATURES IN HORSES, ETC. 171

Zebra's face. We have seen that white, black, and tan are *interchangeable*. Dark Horses have either white or black blazes; and dun and roan Horses have sometimes conspicuously black blazes.

This interchange of colour—general or partial—occurs in various other animals. The black and tan Dog turns into tan and white; the Dalmatian Dog is white, spotted black; certain Deer are of a tan colour, spotted white; and the stripped American Marmot is of a dark colour, with strings of white spots.

Besides Horses there are many other animals that have black blazes on their face, such as several species of Antelope, several kinds of Deer, Ibex, Goats,[1] etc., and the Bonté Bok Antelope (*Damalis Pycarga*) has a white blaze. Cattle have often a white star on the forehead, which we might consider, as in the Horse, a *vestige* of the white blaze.

We may perhaps infer that the black blaze is a fusion of the Zebra's face-stripes, as we have before inferred that certain blotches and stripes on animals are fusions of isolated spots, but why a black blaze should turn into a white blaze, or tan feet should turn into white feet, as in the Dog, is more than I can tell. It would appear to be caused by a sort of atomic 'conjuring' of the nerve-centres hidden from our view even through the most powerful microscopes. Anyhow, we have ceased to think of supernatural causes for all these phenomena. There must be some natural causes for them, although at present we do not know them.

All I know is that white, black, and tan colours in mammals are, as I said, interchangeable, and therefore I surmise that the Horse's white blaze may have been originally black, and that in some varieties it changed into white, and has for some reason become more persistent in this colour.

Whether the blaze of the Horse may have any relation to the

[1] The male *Bharal* (*Ovis Nahura*) has a fine black blaze.

insertion of the horns in some ancestral Rhinoceros is a question difficult to answer. We have seen that in certain animals skin armour has left certain marks when it disappeared, which are liable to change of colour, and not impossibly the blaze on the Horse's forehead may be vestigial of ancestral frontal and nose armour.

It might lead to a certain amount of confusion to compare the Horse now with a Rhinoceros, and then with a ruminant, but Professor Alleyne Nicholson, in his *Swiney Lectures* of 1893, told us that if we go back far enough in time we shall find that animals which are now so distinct as ruminants and carnivora were in remote times mixed up in *one* animal, with features that had something of *both* the branches into which that remote animal eventually bifurcated.

According to Professor Weismann's doctrine, the object of sexual propagation is to mix up the separate descents of one individual with the separate descents of another individual, with the view of rendering the germ plasm more plastic and variable, so that it may the better provide the raw material of variations in organisms for the factory of adaptation by natural selection. We know that the characters of two distinct species, even of two distinct genera, were initially mixed up in one individual.

Well, in bygone ages, it is very probable that animals which were largely differentiated in certain anatomical features were not so differentiated *physiologically* as to prevent their being mixed by sexual mating. If this be admitted, it would follow that what we now would call distinct species, might then have freely intermarried, and procreated what we now would call *hybrids*. For instance, a mixture of hornless, one-horned, and two-horned Rhinoceroses, if they happened to come together, might have interbred and become mixed. In past ages this may have been a frequent occurrence. So with Antelopes, Horses, etc.

INTERESTING FEATURES IN HORSES, ETC. 173

Some one might say that intermixture would eventually have produced an *average* structure, and all differences would be suppressed. But this does not appear likely in all cases, for we know that six-fingered people, although diluted with normal blood, have rather *strengthened* than suppressed the monstrosity.

Such intermixtures are not at all imaginary, for we know that the Pheasant will procreate with the common fowl, and different species of Pheasants will interbreed. And among plants, species so distinct anatomically as a *Lælia* and a *Sophronitis*, a *Cereus* and *Phyllocactus*, and a *Gesnera* and a *Gloxinia*, have been successfully mated.[1]

The application of all this to the Horse is that in our modern Horse we may have not impossibly the convergence of, not only two distinct races of Rhinoceros—the one and the two horned; but possibly also an intermixture of blood derived from some ancestor of the Giraffe, and also of other ruminants,[2] when they were not so differentiated as they are now.

If the reader should carry away the notion that I consider there were once Horses with horns between their eyes and on their noses, he will carry away a very wrong idea of what I mean.

The Horse is a very specialised recent elaboration, as his feet testify, from some remote and more generalised form, which was the raw material out of which various kinds of mammals, such as Giraffes, and other ruminants, Rhinoceroses and Horses have been evolved. Horses or Horse-like animals may never have had horns homologous with those of the Rhinoceros, but their ancestors may have had them, and in the existing Horse, all that there is to tell the tale of ancestral mesial horns may not impossibly be these

[1] At the meeting of the Royal Horticultural Society on the 8th May 1894, Messrs. Veitch and Son showed a hybrid (*Gloxinia* '*Brilliant*') between a Gesnera and a Gloxinia.

[2] In the tuft of long hair over the nose of the white-tailed *Gnu*, we may have the *dissociated* hairs of which the Rhinoceros horn is said to be made up.

frontal and face marks for which I have been endeavouring to discover an explanation. The same nerve-centre *habit of action* which, in ancestral forms, gave rise to horns, now that the Horse no longer has horns, may display itself by producing a change of pigment of the skin in the places once occupied by horns.

There is another feature in Horses for which I have been endeavouring to discover an explanation.

When Horses change colour, it is only down to the ankle and wrist that they usually do so; only rarely does the body colour continue to the hoof; so that we see bay, dun, roan, and dark grey Horses with their hands and feet black; strawberry roans with these parts bay or chestnut;[1] and we see Horses of various colours and even black Horses, with *white* hands and feet. What can be the cause of this? Why should there be so frequently such a distinction between the coloration of the hands and feet of Horses and that of their bodies? It is only in pure whites, pure chestnuts, pure blacks, that the hands and feet are usually of the same colour as the body, but in the duns, or sponge-coloured horses, the black hands and feet form a strange contrast to the colour of the body.

If we make investigations among other animals in the Natural History Museum we find that the Ruffed Lemur has long hair down to its ankles and wrists, and on its hands and feet the hair becomes short, and of a different colour, so that this animal looks as if it had jet black gloves and socks.

The Alpaca has a brown body with long hair, and then suddenly its hands and feet become short-haired and jet black.

The Gaur (Bibos gaurus) and its congener the Javan Ox, have white hands and feet, although the former has a black and the latter a tan or dun-coloured body; so also the Gayal (*Bos frontalis*).

---

[1] Many roans when clipped are of a blackish grey, and strawberry roans either bay or chestnut.

INTERESTING FEATURES IN HORSES, ETC. 175

The Inyala Antelope (*Tragelaphus angasii*) has a roan body and tan hands and feet.

The Waterbuck (*Kobus ellipsiprymnus*) is roan, with brown hands and feet.

Then the 'Bonté bok' (*Damalis pycarga*) is still more interesting, for it has a white blaze on its forehead, like that of the Horse, and also white hands and feet.

I do not think that the doctrine of natural selection will help us much in deciphering these hieroglyphics.

The old Dutch Horse with white hands and feet and forehead blaze, which was introduced into England and much esteemed, may explain why there are so many Horses with these features in England; but it does not explain in the least why the old Dutch Horse had these features; or why, when the black Horse changes into dun, he gets *black* hands and feet, instead of white, and often also a black blaze instead of a white one.

Is there then no explanation of these contrasts of colours, such for instance as those of a *black* Horse, with *white* hands and feet?

It will perhaps be more convenient to take a broader view of these contrasted colours in certain parts of animals.

As I have said, the Horse has frequently its hands and feet differently coloured from the rest of the body; the Dog has frequently tan-coloured hands and feet, and a black body, or white hands and feet and a tan-coloured body; I have seen a white Goat with black hands and feet. But besides these contrasts of the hands and feet, there are others which we find in various animals. The lips, up to a certain point round the mouth, a circle round the eyes, a circle round the base of the ears, and a space round the vent, are all frequently of a colour which forms a great contrast with the surrounding ground.

What is there in *common* with all these parts? The reply is, great *mobility*. If it be true that all mammals descended from armour-plated ancestors, it must be evident that the parts mentioned were the *earliest* to get rid of their armour in order that their struggle for existence might be facilitated. If the hands and feet were covered with hard scales, as in the turtle, it is evident they could not be used to run and grasp; if the eyes were surrounded by armour-plates, as in the Ganoid fishes, it is evident that sphincter muscles, to close the eyelids and protect the cornea, could not be developed; if the base of the ears were not sufficiently mobile to allow of the ear being easily directed towards the source of sound, the muscles that perform that function could not be developed; so of the lips and vent. All these parts are highly mobile, and must have lost their armour-plating *early*, in order to admit of that mobility, and hence we see that these parts had ample time to *change* their coloration *long before* the body got rid of its carapace, or body-armour.

A number of animals show a complete or partial circle round the eye, such as the Ocelot of Fig. 19 (*a*). And also many Cats, such as the Caffre Cat, the Pampas Cat, the Genet, the Binturong, and several others;[1] and so have some tame Rabbits. Then a number of animals have a spot, emphasised by contrast of colour, over the eye like that of the black and tan dog;[2] such as the European Lynx, the Caracal, the Clouded Leopard, and others. In many variations of the black and tan Dog, one often meets with individuals which have a complete tan-coloured circle round the eye.

[1] See *Royal Natural History*, vol. i.
[2] Mr. Worthington G. Smith states that, 'The spots are by no means always tan; a black Dog will sometimes have them white, and a white Dog black. I have a white and tan Fox Terrier in which the spots are very eye-like and jet black; in a brown Bull-pug of mine the spots are also black.' (See *Nature* of 15th November 1894.)

INTERESTING FEATURES IN HORSES, ETC. 177

One would not suspect that the Horse came under the same category, but I have seen a dark-bay Horse which had a semi-circle below the eyes, and a spot above them, both of a tan colour, and quite distinct from the general colour of the face. All these seem to be no other than *vestiges* of ancestral circles round the eyes, contrasted with the general colour by a different pigment, and in all probability derived from the cause I have hinted, viz., *early loss of armour* round the eye, to admit of closure of the eyelids for the protection of the delicate structures of the eye.

It must be obvious that in the transition of sea-animals without eyelids, to land animals which moved about in dense forests, it was of importance for them to acquire an unarmoured circle round the eye, and the faculty of closing their eyelids to protect the cornea from injury; and especially at night it was of importance that long hairs should grow on the unarmoured circle round the eye such as we see in the white-tailed *Gnu*, to warn them of the vicinity of a branch, or a leaf, or other object, *before* the surface of the cornea came in contact with it, and thus enable them to close the eyelids in time.

And so of the other parts in which mobility was of great importance—the hands and feet, the base of the ears, the lips and the vent. In some specimens of heads of the Black Buck which I have seen, there was a distinct white circle round the base of the ears. A large number of dark-coloured Horses and Asses have a light-coloured mouth. The Cyprian wild Sheep[1] has a fine and striking white muzzle, beautifully contrasted with the dark face. Who would have thought that the origin of this white muzzle was, in all probability, the unarmoured muzzle of some animal not unlike a Pangolin? All the coloration of this Sheep is indicative

[1] *Royal Natural History*, vol. ii. p. 225.

of ancestral armour which had passed through stages of *partial disarmour*.

I have seen a *white* Pony which had the mouth, circles round his eyes, and circlets just above the four hoofs, *all* of a yellow or rather golden-bay colour. It is evident that in these parts there was some *difference* of innervation—a remote ancestral habit of the nerve-centres—which prevented a *total* albinism of the skin.

Of course, in the great battle of life, and the many adaptations to surroundings to which animals have been subjected, and in the fancies of breeders of domestic animals which have revolutionised the coloration of a number of animals, it is not to be expected that they could long retain the coloration of their unarmoured parts *unmodified*. We therefore now see mere *vestiges* of ancestral features. It is only occasionally, by a sort of reversion, that we are let into the secrets of what may have obtained in remote times. The very fact that so many *different* kinds of animals have these features is sufficient indication of their *antiquity*.

As I said, there is no hard and fast *fixity* of coloration in these parts; there is not only an interchangeableness of black into tan, or shades thereof, and of tan into white, but also an intermingling of the coloration of contiguous surfaces, as well as a dwindling, in many cases of the markedly coloured surface. For instance, we see the white hands and feet dwindling to the last phalanges in certain Cats and Dogs, and in Horses to a mere vestige of white just above the hoof. In the Horse all four limbs may be white, or only three, two, or even only one, may be white, so that the contrasted colours often merge into the *general* coloration. The curious thing is that in the Horse there would appear to be some correlation between the mane and tail and the four feet, for all these are often of the same contrasted colour, either black or white, or shades of tan, when the body is of a totally different colour.

It is very curious that the wild cattle of Cadzow Park[1] should have *black* muzzles, ears, and front parts of hands and feet, while the rest of their body is *white*. It would seem that those parts had originally a specially different function from the rest of the body, which is now of a contrasted colour, while in the *Gaur* the body is black, the muzzle pale, and the hands and feet *white*.

Then in the same work[2] it is stated of the wild cattle of the Somerford Park breed, that 'the colour is pure white; the ears, rims of the eyes, muzzle, and hoofs being quite black. Like all other herds of the forest breed they have a tendency to produce black spots on the neck, sides and legs.'

Surely this contrast of certain parts with the general coloration, and this tendency to produce spots, cannot be the result of pure accident. It seems that in the nerve-centres the 'memory' of ancestral dermal conditions is now aroused, and now in abeyance, and so the results on the skin are now ancestral, and now aberrant.

In speculating on the causes of these interesting phenomena, we should not be led away by the bias of a theory, and shut our eyes to any side-lights that may turn up in the course of our research. I shall therefore put on record the following fact, so that the reader may make of it what he can.

In the Tring Museum there is a fine specimen of Chapman's Zebra. It has the lower parts of its limbs for several inches above the hoofs *wholly black*; while Grevy's Zebra close by has those same parts *fully striped*. It would appear that the black of the former may be a fusion of the stripes of the latter; but at the same time it would show that there was a *difference* of innervation there, which *caused* this fusion.

Among the extinct Ganoid fishes there were several species

[1] See *Royal Natural History*, vol. ii. p. 164.   [2] *Ibid.* p. 167.

with a circle of plates round the eyes, as shown in Fig. 76, (*a*) and (*b*), and others in which the eyes were mere holes in a solid one-piece head armour as in (*c*).[1]

In their descendants, the carapaced land animals, it may have been this circle of plates round the eyes which was early got

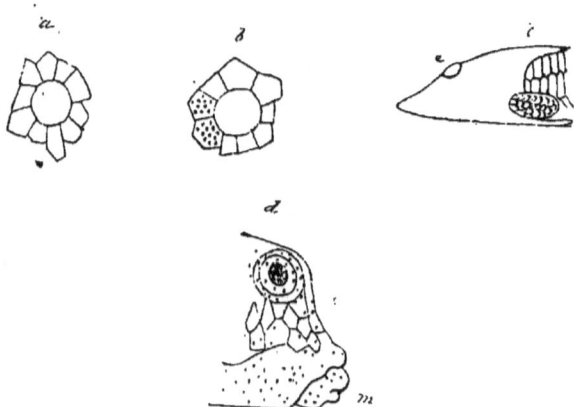

FIG. 76.—(*a*) Circle of bone-plates round the eye, *Lepidotus maximus*, p. 986; (*b*) Tuberculated bone-plates round the eye of *Dapidius pholidotus*, fig. 922; (*c*) Solid bone mask of *Cephalaspis Lyelli* with an eye-hole at (*e*), p. 961 (all three Ganoid fishes from *Manual of Palæontology*, by Nicholson and Lydekker, vol. ii.); (*d*) is the head of *Ostracion punctatus*, an armour-plated modern fish, with its unarmoured lips at (*m*) from pl. 181, *Fishes of India*, by Francis Day, pt. iv.

rid of to enable them to close their eyes, if necessary, while moving among bushes, for the protection of their eyeballs. Then there is that curious modern fish called *Ostracion*, which is encased in hexagonal bone-plates all but the lips; the fins and tail also have no plates round their base. Fig. 76 (*d*) gives a sketch of its head.

[1] See Figs. 898, 922, and 926 of *Manual of Palæontology*, Nicholson and Lydekker.

INTERESTING FEATURES IN HORSES, ETC.   181

The still remaining spots on the lips and round the eyes suggest to my mind that these parts also, in some ancestral form, were encased in plates. The anal sphincter-muscle may similarly have required the absence of calcareous deposit.

Then I should say that the early loss of armour round the eye, round the mouth and vent, and also from the hands and feet, parts acquiring great mobility in the struggle for life, is the reason why, in the higher animals, these parts are so often *contrasted in colour*. The early loss of armour gave those parts the opportunity of becoming variously pigmented *long before* the rest of the body had become uncarapaced. The base of the ears, and the lower parts, such as those of the neck, abdomen, and tail, would follow the same rule.

That is, they lost their armour, replacing it by hair, long before the other parts of the body lost theirs; and therefore they had time to alter their coloration before the parts with which they are contrasted had become armourless.

As shown in another place, we see frequent examples of alterations in hair colour on parts occupied by hairless skin-structures which became subsequently suppressed. The most convincing examples of this are the tufts of differently coloured hair which have replaced the callosities on the legs of certain ruminants (see Fig. 85). From a minute study of animal coloration, it seems clear that small portions of the skin can alter *independently* of adjacent surfaces, or without the coloration of the whole surface becoming altered.

But if we are anxious of being thoroughly convinced that this ancestral loss of armour in certain parts is a *vera causa* of the contrasts of pigments in their existing descendants, we should visit the Pangolins in the Natural History Museum.

There we find that the Malayan Pangolin (*Manis Javanica*), and the long-eared Pangolin (*Manis Aurita*), have no scales on the under surfaces of the head, neck, chest and abdomen, and also on the inner aspects of the four legs.

Then the yellow-bellied Pangolin (*Manis Tricuspis*), as well as others, has a muzzle *wholly destitute* of scales, and its under surfaces and paws are hairy instead of scaly. The soles of its hind feet from the heel to the toes are also *bare*; and this latter fact will give us a hint why, in certain descendants of armoured animals, such as certain Cats and Dogs, this very surface from heel to toes in these *digitigrade* descendants is *black*, and contrasted with the pigment of their adjacent surfaces.

It is true that these Pangolins still retain scales on the under surface of their huge tails, but if we search further we find that in the little Ant-eater close by (*Cycloturus Didactylus*) the under surface of the tail is hairless, so is that of Horses; and this may be the reason why, in the innumerable changes of skin coloration which have occurred in mammals, the under surface of the tail is so often contrasted in colour with the upper surface. Indeed, the under surface of the tail is only a continuation of the under surface of the body from the muzzle to the tip of the tail, and the white tip of the tail we see in so many animals may be only a *vestige* of a more extended whiteness of the under surfaces.

Then do you mean to say that Dogs, Cats, Asses, Antelopes, etc., which have contrasted upper and under surfaces, are descendants of Pangolins?

I mean to say no such thing. The Pangolins, as well as the Armadillos, and these other mammals mentioned, are descendants of a much remoter stock which had their bodies wholly armoured up and down. Armour was first suppressed on the under surfaces, and gradually replaced by hair, and the Pangolins and Armadillos are *survivals* of that state of things. Other descendants of the

INTERESTING FEATURES IN HORSES, ETC. 183

same stock, and much later on, wholly lost their armour, and acquired in their hair contrasted colours *instead*.

So that in existing mammals we seem to have indications of *four* distinct stages of evolution—

(*a*). Mammals which were wholly armoured, somewhat similar to, say, the existing Crocodiles ;

(*b*). Mammals which were half-armoured, with hair on the lower, and plates or scales on the upper surfaces, as in the existing Pangolins and Armadillos ;

(*c*). Mammals which had wholly lost their armour, but retained indications of their ancestral partial armour in their rosettes and in their contrasts of coloration ; and

(*d*). Mammals which have lost or are losing every indication of ancestral armour, such as the Puma, the Lion, the domestic self-coloured Horse, the domestic self-coloured Cats, Dogs, etc

The white circle round the eyes and the white muzzle are well shown in the white-nosed Coati of Central America. Its ringed tail sufficiently indicates its descent from an armour-plated ancestor, if the reader has been following this interesting discussion. But the circle round the eyes is best seen in the parti-coloured Bear of Eastern Tibet.[1] This particular animal is very strangely marked. It is white, with black rings round the eyes, and black ears, while the shoulders are marked with a transverse black band broadening towards the fore-limb. The under surface and legs are black.

The writer who describes it says : 'Without knowing more of its general surroundings, it is difficult to imagine the object of such a staring contrast.'

Perhaps the coloration of this Bear has nothing to do with surroundings as a cause of its 'staring contrast,' but may be a vestige of an *incomplete* ancestral carapace, viz., that of a neck and head

[1] *Royal Natural History*, vol. ii. p. 33.

shield, and a body and haunch shield, the shoulders and limbs, and circles round the eyes, having *early* lost their armour; while the Cape Ratel, and the common Badger may have descended from ancestors which had a more complete armour.

If the reader will turn to Fig. 79, of *Rhinoceros Sondaicus*, he may perhaps be struck with the possible homology of its scapular shield with the staring scapular black mark of this parti-coloured Bear! Even the fold in the groin of the Rhinoceros would seem to be imitated by the black pigment in the Bear's groin—and its white head and neck are suggestive enough of the neck and head armour of the Rhinoceros.

One would be tempted to infer from all this that the remote ancestor of this Bear lost its scapular armour long before it lost its other carapacial armour; and we know that the extinct *Polacanthus* had only a *partial* carapace on its lumbar and pelvic regions.

In this and similar speculations allowance should naturally be made for the transition of solid plate-armour to its copy in hair, and for any modification the Bear-descendant may have undergone in its coloration through the ages.

For instance, *Ursus Americanus* is wholly black, excepting its muzzle, which is of a tawny yellow. The *Tayra* of the Weasel family is usually of a dark brown colour, but occasionally white specimens are met with, their muzzle, ears, and feet, however, *remaining dark*. Of course I do not in any way mean that this bear descended from *Rhinoceros Sondaicus*, any more than I mean that the Giraffe descended from a Zebu; all I mean is, that, in tracing their origin backwards, the skin characters of certain mammals indicate that they converge towards an armour-plated ancestry.

Professor Huxley says: 'In former periods of the world's history there were animals which overstepped the bounds of existing

INTERESTING FEATURES IN HORSES, ETC. 185

groups, and tended to merge them into larger assemblages. They show that animal organisation is more flexible than our knowledge of recent forms might have led us to believe; and that many structural permutations and combinations [I would take the liberty of adding skin features],[1] of which the present world gives us no indication, may nevertheless have existed.

'But it by no means follows, because the *Palæotherium* has much in common with the Horse, on the one hand, and with the Rhinoceros on the other, that it is the intermediate form through which Rhinoceroses have passed to become Horses, or *vice-versâ*; on the contrary, any such supposition would certainly be erroneous.'[2]

As I said, there is no fixity of coloration round the eye or other part, as it may dwindle and disappear altogether, or it may invade adjacent parts. Those animals which have a complete circle round the eye, whether white, black, or tan-coloured, are survivals of a stage which at one time may have been general, and the spot over the eye of the black and tan Dog may be only a remnant of a former complete circle round the eye.

. On one occasion I saw a jet black Persian Cat. The tips of all its four feet were white; it had a white patch on its chest, and a little white on its abdomen; its whiskers and eye hairs were white. There was no contrast of colour round the eye and round the lips—both being jet black; but these white whiskers and eye hairs were seemingly the *only* remnants of surfaces differently coloured from the body.

There are other curious marks in some mammals which are not so easily accounted for; such as the white tip of the tail in many Dogs to which I have already alluded, and a white patch which one sees so frequently on the chest of Dogs. There can hardly be much

[1] The parenthesis is mine.
[2] 'Science and Hebrew Tradition.'—*Lectures on Evolution*, p. 102.

doubt that these two marks are inherited from the Wolf. When in India, I examined a large number of very young Wolf-cubs, which had their eyes still closed. They were usually of a dark chocolate colour, and a large proportion of them had a white mark on their chests, and a white tip to their tails. Probably in the Wolf-cubs these are only vestiges—remnants, as I said—of a more extended light-coloured under surface. For, as the Wolf-cubs grow, the white becomes blended with the general fawn or tan colour of the under surface, and eventually totally disappears, while the dark colour of the back turns into an intimate mixture of black and tan resembling that of certain Collies whose black upper surface had changed into a mixture of black and tan.

As regards the white tip of the tail of certain mammals, there are some curious phenomena connected with tips—tips of tails, tips of ears, and tips of feet—which would point to some difference in the innervation of tips. As I said, white and black are interchangeable. There are many mammals which have black tips to their tails, and this, in allied or other individuals, may change to white. The Arctic Hare in its summer clothing is brown, with black tips to its ears; and the ermine is also brown with half the tip end of its tail black. When these two animals get their snow-white winter clothing, the tips of the ears of the one, and the tip of the tail of the other, remain *black!* I have seen a number of skins of the Arctic Fox in the furriers' shops, and many of them had a tuft of black hairs on the tips of their tails. There are two coloured pictures of the red common Fox in the *Royal Natural History*. One has a black tip to its tail, and the other a white tip.

In this connection I would note a curious mark I have seen on two sister Cats of the tabby variety; (*a*) had pale tan hands and feet, and on the hind limbs, from the heel to the toes, the *back* surface was *black*; while (*b*) had white hands and feet, and only a *vestige* of black near the heel. Not improbably this black surface on the back part of the hind limbs, as I have already hinted, may indicate that some ancestral form was *plantigrade*, and in changing to a digitigrade descendant, a record was left of its ancestral plantigradeness in this *contrast of colour*. I have seen the same black mark on the hind legs of Collie Dogs; and if the reader will turn to the Serval of Fig. 17, he will see that the same surfaces are black, which I take to be a *colour*-vestige of an ancestral *plantigrade* surface. Then, curiously enough, we find that the hind edge of the metatarsus of *Tragulus*—a pigmy Chevrotain, allied to Pigs—is *naked and callous*—a veritable plantar surface, without hair, inherited from some plantigrade ancestor, not unlike a Pangolin, whose soles are bare. In the evolution of digitigrade carnivora and ruminants from plantigrade ancestors, with bare soles, this surface became elevated above the ground, and in some became hairy, hence the contrast of colour there in certain Cats, Dogs and the Serval, while in the *Tragulus* it remained *bare*.

IS NATURAL SELECTION THE SOLE FACTOR
IN THE COLORATION OF MAMMALS?

'If one wishes to realise the wonders of natural selection, he or she should go to the Natural History Museum, in South Kensington, and inspect the Homopterous insect called *Flatoides dealbatus*, and think how much decimation of generations it must have required to make that insect's wings *indistinguishable* from the Lichens on the bark of trees.'

PART VII

IS NATURAL SELECTION THE SOLE FACTOR IN THE COLORATION OF MAMMALS?

IN studying this intricate subject the reader should not mind a little repetition, as it may be needed to *emphasise* certain points.

In searching for a cause of the spotting of certain mammals, it is not sufficient to say, in a vague way, that it has been brought about by 'natural selection,' through which it was made to harmonise with its surroundings,[1] for there are in existence certain facts which do not seem in the least to admit of such an explanation.

In the Science and Art Museum of Edinburgh, there is a specimen of the Great Armadillo (*Priodontes maximus*). Its abdomen is studded with *separate* rosettes, composed of very minute plates ; and here I would wish again to emphasise the fact that the hind feet of this Armadillo are *ungulate*, in the fashion of those of the Rhinoceros and American Tapir, and very different from its forefeet, which have long claws.

Curiously enough, the Peba Armadillo (*Tatusia Pebâ*)[2] has no

---

[1] Where the black and spotted Jaguars are common, they can hardly *both* harmonise with their surroundings. Dr. Wallace (*Travels on the Amazon*, p. 317) says: 'In some localities the black Jaguar is unknown, while in others it is as abundant as the ocellated variety.' On the other hand Mr. Saunderson says that the black Leopard in India is confined to heavy forest tracts, while the ocellated variety frequents open country and rocky localities.

[2] Science and Art Museum, Edinburgh.

plates proper on its abdominal surface, but, on its chest and between its hind legs, it has the merest varnish of former scales. They are not unlike in nature to the black dabs on the lips of *Serranus Gigas*, after the scales of the latter have thinned out into nothing. The spots on the middle part of the abdomen of this *Tatusia* are less distinct.

Armour on the abdominal surface of the Armadillo would evidently have been inconvenient to an animal that rolls itself up into a ball, and shows its armoured back to the enemy. It seems that the Tatusia lost its abdominal plates, because they had become not only useless, but inconvenient, or perhaps from some other cause ; it, however, retained their vestiges as *spots*. The alternative supposition would be that the Tatusia took to rolling itself up, when it began to lose the stiff plating on its abdominal surface, while the dorsal armour divided itself into moveable sections, which admitted of its rolling itself up.

However this may have been, it is clear that the Armadillo descended from an ancestor which was armoured both dorsally and ventrally, somewhat in the way shown in Fig. 73. It is stated that certain Glyptodonts had an armour on their ventral side, which suggests that of the Turtle.

The existing Crocodiles have retained their bony dorsal plating, while the ventral armour is made up of hard horny plates. In the Crocodile, each plate is surrounded by skin, and its movements are much freer than could have been those of the Glyptodon.

I would now ask—what business has the Peba Armadillo with spots on its under surface, supposing them to have been originated by a simple process of *natural selection*? The spots on the abdominal surface could in no sense harmonise with their surroundings, and so become protective, for, when this Armadillo is unfolded, and moving about, its abdomen is close to the ground, and

FACTORS IN THE COLORATION OF ANIMALS 193

the spots hardly visible, while, when the animal is rolled up into a ball, the spots are wholly *invisible*.

In the case of Jaguars and Leopards, which are also spotted on the abdomen, it might be argued that, as these animals may sometimes climb up trees, their abdominal surface, looked at from below, would harmonise with the lights and shades of the foliage;

FIG. 73.—One of the Crocodiles in the Zoological Gardens, from a photograph.

and in the case of a Tiger, which has a striped abdomen, we might say that, if it were lying down among the long grass, in which it is often found, its abdomen would also harmonise with the lights and shades of the grass stems. But no such excuse can be found for the spotting on the abdomen of the Armadillo, nevertheless the spots on its lower surface are there!

To my mind they can only be explained as impressions—

194 STUDIES IN THE EVOLUTION OF ANIMALS

vestiges—left on the skin by plates in some ancestor, in which plating was useful there, such as in Ganoid fishes and other animals, like the Crocodiles, which pass a part of their lives in water.

Then we cannot say that the dappled Horse, and the Zebra, and the Giraffe, climb up trees, and the one requires its spots, and the other its stripes, to harmonise with its surroundings; yet they are abundantly spotted, or striped, on their abdominal surfaces.

The spotting, and of course the striping is a derivation from it, seems to be a survival on parts of the skin where *once* ancestral plates *were*. To the evolutionary philosopher they are of great importance, for they indicate a link between skin-*plating*, and skin-*spotting*.

It may be interesting to note that the embryo Armadillo (*Tatusia hybrida*),[1] shows plates on its abdominal region also!

It would appear that through heredity, not only bones, muscles, etc., are transmitted, but also skin impressions, as if they were eternal photographs of former plating.

Mr. R. Le Gallienne has written:[2] 'Nature ruthlessly tears up her replicas age after age, but she is slow to destroy the plates. Her lovely forms are all safely housed in her memory, and beauty and goodness sleep securely in her heart, in spite of all the arrows of death.'

In other words, applied to our subject, this would mean that the forms and colorations of animals, when once established, are slow in becoming altered; that although the individual is being continually destroyed, the mould that produced it is more permanent than many think, and that provided the individual has

---

[1] Professor Parker's *Mammalian Descent*, p. 94.
[2] *The Religion of a Literary Man*, p. 50.

FACTORS IN THE COLORATION OF ANIMALS 195

had opportunities of leaving descendants, the original characters will somehow sooner or later reappear.

Then, from the point of view of natural selection, there is no good reason why there should be such a sharp contrast of pigments on the flank of the black Buck, and many other animals; but, from the point of view of a *hereditary impression*, there seems very good reason why the pigments which picture the separation of the formerly armoured and unarmoured surfaces should be contrasted. It seems to represent the line of division which we now see between the back and abdomen of the Armadillo (Fig. 71), and of some Pangolins. The white or *differently* coloured abdomen would then mean that the abdominal surface lost its armour, and may be also its spotting, *long before* its ancestral forms lost their carapace.

This theory of the origin of contrasted colours is based on the early disappearance of armour from certain parts of the surface while it continued to a later period on certain others; and the contrast of pigments, when finally the armour wholly disappeared, represents the two stages. It does not attempt to account for *every* bit of colour, but only for those colorations which appear to be formed on a *plan*.

To put this idea more clearly before the reader—The repetition through ages of the nerve-centre action which resulted in the formation of dermal bone-rosettes, left, as it were, a *memory* in the nerve-centre. This memory, when the dermal plates could no longer be produced, externated itself by producing only *picture*-rosettes, as we see them in the Jaguar. When even the memory of separate rosettes had been effaced, it still continued to externate itself by producing a contrasted pigmentation of ventral and dorsal surfaces; that is, of the sites of the *earlier* and the *later* loss of armour.

Among existing animals we have survivals of all the stages that mammals have gone through in their evolution. We have animals which are armoured above and below; animals which are armoured only above; animals which are spotted above and below; animals which are spotted only above; others which are darker above and lighter below, with modifications in other parts; and finally animals which have lost every trace of ancestral armour, and have become self-coloured throughout.

All these changes on the surfaces, it would appear, have their 'molecular equivalents' in the nerve-centres; and when modifications, from whatever cause, occur in these, their representatives on the surface undergo a simultaneous modification.

What wonder is it that some sort of picture of bygone states should now be produced, when we know that every sensory impression leaves a record in the grey matter which we call *memory*?

Now how did natural selection come in, in the case of Jaguars and Leopards? These carnivora are expert climbers and pass a part of their life on trees. Their skin-rosetting harmonises well with the surrounding lights and shades thrown by the foliage, and so this curious coloration has been *maintained* by natural selection, and not *made* by it, while in the case of the adult Puma and Lion the same process may have *selected away* the picture-rosettes.

Of course the surviving Leopards that frequented trees did not go there because they *thought* the lights and shades of the leaves would protect them. They frequented trees for *other* purposes, and found that *there* they were *protected*—that is, there they could *secure prey* more easily than on the plain or among bushes; and so the Cats with these peculiar markings have been *preserved*. However undeveloped the brain of a wild or domestic animal may be, it knows very well *where* it can get food easily.

To try to account for every bit of coloration in existing

FACTORS IN THE COLORATION OF ANIMALS 197

animals by 'natural selection' would be an idle task. No amount of natural selection, for instance, would account for the fact of so many mammals having ringed tails and plain bodies (see Appendix E. Nos. 4, 7, 11, etc.); for we cannot for one moment suppose that it was the *tail alone* that required protection! In such cases we can only fall back on the supposition that, for some reason unknown, the tail has been slower in losing the modified rosettes than the body.

As happens with most theories, to pursue this one further would entangle it in difficulties, as so many modifications in the coloration of mammals have occurred since their ancestors wholly lost their armour. It is only by an occasional survival here and there that some sort of concatenation can be conjectured to link the armoured stage of mammals with the existing stage, which in most cases is totally different.

The upshot of this discussion is that we come to the conclusion that the rosettes of Leopards are the inherited *imprints* of armour-clad ancestors, and *not* the result of natural selection by slow and useful modifications; and that their strange coloration has been *preserved* by natural selection because it may have harmonised with the surroundings that these animals frequented.

PROBABLE CAUSE OF THE LOSS OF THE CALCAREOUS ARMOUR IN MAMMALS

'During sudden changes in the home or feeding-ground of animals, the dilemma has again and again been *adaptation* or *extinction*; in many cases nothing short of metamorphosis has saved them from death and kept them alive in famines.'

PARKER'S *Mammalian Descent*, p. 18.

PART VIII

PROBABLE CAUSE OF THE LOSS OF THE CALCAREOUS ARMOUR IN MAMMALS

BEFORE entering into this discussion, it may serve some purpose to go over some of the points discussed in the foregoing pages.

A glance at the three Figures 4 to 7 shows that the groups of spots on the abdominal surfaces are much more altered than those on the back and flank. This, in my opinion, as I have often stated, has resulted from the abdominal armour having been got rid of *before* the dorsal armour, so that the abdominal rosettes had time to alter before the dorsal pigment-rosettes had come into existence, because the dorsal region was still covered by a *carapace*. In other words, the ventral spots of the Leopards are much more *ancient*, and consequently more modified than the dorsal rosettes.

One might suppose the process of dispensing with armour-plating was something after this fashion :—The remotest vertebrate ancestor was plated top and bottom, like the Ganoid fishes and the Crocodiles.

Mr. Alfred Russel Wallace[1] says that, in skinning the *Jacaré* —an Alligator six feet long—'the scales of the belly could only be cut by heavy blows with a hammer on a large knife,' and that he 'was obliged to borrow a drill to make the holes to sew up the skin,' after stuffing it.

Then the complete rachitis of the skin occurred first on the

[1] *Travels on the Amazon* (Rio Negro), p. 219.

ventral surface, as the part least needing armour, especially in land animals, and a type of mammal was reached which is now represented by the surviving Armadillos and Pangolins.

When the armour disappeared from the ventral surface it left behind the imprints of armour-rosettes in the shape of pigment-rosettes and spots.

Later on, for some reason which I shall come to in a little while, a type of animal was reached in which the dorsal armour also began to dwindle. For instance, *Thoracophorus* and *Mylodon Darwinii* had separate scutes or plates on their carapace, and others, such as *Carioderma*, had only *vestiges* of plates.

We see a similar diminution and scattering of dermal plates, and probably from the same cause, in some Trygons and in *Echinorhinus spinosus* among fishes.

This dermal rachitis was evidently *progressive*, and eventually a stage was reached in which the whole armour disappeared, leaving only imprints of its former bone-rosettes.

Thus we come to the armourless ancestral forms of the modern Jaguar and Leopards, with only *pictures* of scutes, with all their modifications, as detailed in the foregoing pages.

' Of course all this descent is a speculation, but it is founded on the facts which I have endeavoured to place before the reader. . I need not add that all these changes did not occur in a *day*.

There are many minute and other forms of spotting and striping among different classes of animals which cannot honestly be said to be of any use whatever for individual protection, although they may be of service for social ends, such as *recognition*, etc. Animals, whose minds are not so much engrossed by fashions and the frivolities of society, may perhaps take far more notice of a few characteristic spots and stripes on their associates than my friend the Lady-Fellow of the Zoological Society could do, who, although she had been to the 'Zoo' hundreds of times, yet had *never seen* the spots on the Lions.

For some reason, spotting appears to endure longest on the legs, and striping on the tail. The spots, stripes, and ring-tails of animals are hieroglyphics which appear to have a profounder meaning, as to the origin of those that bear them—and possibly also to our own origin—than may have been suspected.

It is difficult to look upon the ringed tail as otherwise than a remnant of a spotted or striped body. As there are numerous animals, such as Racoons, Coatis, and many others, which have only this remaining vestige of a spotted ancestry, it will be seen how many animals must have lost their body spotting.

Both for observation among animals in a state of nature, and among those under domestication, and also for experiments in cross-breeding, there is an almost unexplored field open to any one who may be interested in evolutionary biology; I mean the study of the changes which occur, from birth to old age, in the markings of the skins of animals. The changes that occur in the pigments of the skin are tell-tales, not only of the organic phases that particular animal has descended through, but also of the atomic changes in the nerve-centres which occur in the life of the individual.

We come now to the discussion regarding the probable cause of the loss of the calcareous armour in mammals.

In the evolutions and revolutions which have occurred among organisms from the beginning of time, one might well ask what has caused the loss of those unwieldy bony carapaces and shields, which, in the case of the Glyptodon, as Professor Parker remarked, ' looked as if made for eternity '?

Of course no one would suppose that plate-armour of that sort was dispensed with all of a sudden, although it is conceivable that some monstrous form, *without* plate-armour, may have occurred all of a sudden, just as hairless animals have suddenly appeared.

In *Chlamydophorus* we have probably a curious illustration of a transition form. The calcareous shield has disappeared, and *hair* has grown in its stead ; and over this the horny covering of the bony carapace still remains. By getting rid of the latter also we would have the finished hairy mammal, with pigment-rosettes on its hair to show its connection with carapaced animals.

But the reader might say this tiny Armadillo *has no spots* on its hairy coat, although it would seem to have recently lost its bony shield. No one would expect the law of spotting to be so unchangeable as not to admit of any deviation. Indeed, in some forms of Lynx we have the back unspotted, while the abdomen is spotted and the short tail ringed.

To return to our question then—What could have caused the gradual disappearance of armour, such as we know to have existed in many extinct mammals, which at one time appear to have been very numerous? It is stated[1] that 'the habitat of the Armadillos extended from Mexico and Texas to Patagonia, and during the Tertiary Period was inhabited by the Glyptodons—gigantic Armadillos whose remains are found abundantly in the bone caves of Brazil.'

It is evident to me that these armoured mammals lost their armour gradually. Some had very thick bony plates, others thin ; some had their scutes commissured, others had them separate ; in others, finally, there were only *vestiges* of bony scutes.

This is what Palæontologists say,[2] under the heading of Glyptodons : 'Carapace has no movable bands, so that the animal could not roll itself up ; . . . the carapace usually has its com-

---

[1] *Encyclopædia Britannica*, vol. ii., 9th edition.
[2] *Manual of Palæontology*, Nicholson and Lydekker, vol. ii. p. 1292.

LOSS OF CALCAREOUS ARMOUR IN MAMMALS 205

ponent scutes united by suture, but in one genus they are separate ; the scutes are moreover usually ornamented with a sculpture which varies in the different genera and species ; but they may be plain or tuberculated. There is usually a ventral buckler (never found in the Armadillos), and the tail is enclosed in a complete bony sheath.'[1]

Then, at p. 1294—' *Thoracophorus* differs from all the foregoing in having the scutes of the carapace separated from one another, and thereby approximates to some of the *Megatheriidæ*, in which there were a number of small ossicles imbedded in the dermis of the dorsal region. A similar condition prevails in *Carioderma* of the Loup-Fork beds of Texas.'

Further, at p. 1299—' The Patagonian *Mylodon Darwinii* (*Gryptotherium*) has numerous small dermal scutes, which do not articulate with one another.'

From this it becomes evident that the group of mammals with scutes or plates on their skin (*Glyptodontidæ* and *Megatheriidæ*) was not only a very large one, but it shows that the loss of plate armour, in some, was gradual. The plates or scutes dwindled into *separate bony vestiges imbedded in the skin.*

Now, to endeavour to answer the question I have put, I can only venture on the following speculation :—

To account for such accumulations of coal as are known to exist, physicists have supposed that at one period of the earth's history there must have been vastly more carbon dioxide (CO^2) in the atmosphere, in order to furnish the carbon which such prodigious vegetation must have needed.

Sir Robert Ball has informed us that the clouds of the

[1] In the Manchester Museum there is a Lizard without a name. It is ticketed 'a *Lacerta* from Lower Egypt.' Its great peculiarity is the tail, which is enclosed in spiked rings, almost exactly like a miniature tail of a Glyptodon. The rest of the body is covered with very minute scales.

photosphere of the sun are no other than pure incandescent carbon. Owing to the intense heat, the carbon is dissevered from all its possible combinations and shines and radiates heat as the coal in our grate does, or as the carbon filament in the electric lamp does.

It is logical therefore to infer that our earth had, at one time, a vast deal of this same material in its atmosphere. Then, as it cooled down, the carbon began again to combine with oxygen [1] to form CO^2. If this be so, it would follow that at one time our atmosphere was choked with CO^2, and all life, as we now know it, must have been impossible till the evolution of low forms of plants undertook to liberate the oxygen and hoard up the carbon in the plant tissues.

Now a large quantity of CO^2 in the air would also mean a large quantity of CO^2 absorbed by the water. And this again would mean a large quantity of *carbonate of lime* in solution in the water. Hence we find all the lower forms of primitive life, even down to seaweeds, mostly encased in calcareous material, forming more or less thick shells of innumerable forms, carapaces, exoskeletons, etc.

It is only necessary to think of the chalk and other limestone rocks,—which are formed of countless myriads of foraminiferous shells,—and of the coral rocks, to have some idea of the vast amount of carbonate of lime which has been withdrawn from the waters of our earth ; for these minute animals and corals could have got the lime they needed only from the lime salts in solution in water. Moreover, the fauna of the seas could obtain the materials of both their inner and outer skeletons only from the same source.

The abstraction of carbon from the atmosphere, and the hoarding of it in coal measures, and the abstraction of the same carbon, in the shape of carbonate of lime, from the waters, and the hoarding

[1] Oxygen may not have been detected in the sun, but we *have* it in our atmosphere.

LOSS OF CALCAREOUS ARMOUR IN MAMMALS 207

of it in lime rocks, must have, in course of time, caused something like a *lime famine* in the process of multiplication of animals, hence the *nudity of a number of animals whose ancestors at one time must have been more or less encased in calcareous armour.*

Lime salts must have been abundant everywhere, not only in water plants and animals, but also in land plants and animals. Lime was largely needed not only for building up of the *exo*-skeletons, but also for building up the *endo*-skeletons of those extinct monsters, the thigh-bone of some of which was taller than a six-foot man, and proportionately massive.

It is therefore not to be wondered at that animals, multiplying indefinitely, may have eventually reached a period in which there was an *insufficient supply of lime to supply all demands.*

The endo-skeleton, if not sufficiently supplied with lime, would become *rachitic.* The animal with a soft endo-skeleton would collapse, and be of no use in the battle of life. So it was the exo-skeleton that had to dispense with the calcareous stiffening, and the animal world may then have begun to experience the pathological phenomenon of animals with a *rachitic skin!*—certainly a great advantage where activity had become a factor in the struggle for life.

Even when land mammals had lost their exo-skeleton, they must have continued to use up incredible amounts of carbonate of lime for building up their endo-skeletons; and when we read of the swarms of ruminants which blackened the plains of Africa and North America, we begin to realise the amount of this lime salt that must have been hoarded in their endo-skeletons. But this salt of lime such countless ruminants could evidently obtain in sufficient quantity from the grasses and trees on which they fed.

And this very carbonate of lime which nourished the grasses

and trees was obtained from the disintegration of the minute shell covering of myriads and myriads of far remoter ancestors. These shell coverings had been hoarded ages ago in the rocks, or on the debris of rocks on which those grasses and trees grew!

In the *Royal Natural History*, vol. ii. p. 193, it is written that, 'of all the quadrupeds that have ever lived upon the earth, probably no other species has ever marshalled such innumerable hosts as those of the American Bison. It would have been as easy to count or to estimate the number of leaves in a forest as to calculate the number of Bison living at any given time during the history of the species previous to 1870. . . . As an instance of these enormous numbers, it appears that, in the early part of the year 1871, Colonel Dodge, when passing through the great herd on the Arkansas, and reckoning that there were some fifteen or twenty individuals to the acre, states, from his own observation, that it was not less than 25 miles wide and 50 miles deep,' and was reckoned to consist of not less than four millions. 'Many writers, at and about the date mentioned, speak of the plains being absolutely black with Bison as far as the eye could reach.' It is even stated that trains were sometimes derailed on the Kansas Pacific Railway when attempting to plough their way through these crowds of Bison.

The reckless destruction of these useful animals, for the sake of their fine skins, has left little more than 'huge stacks of skulls piled up at many of the railway stations awaiting transport,' perhaps to make bone-black for refining sugar. Such is the irony of nature!

Mr. Hutchinson,[1] quoting from Darwin, on the American continent, says—'Formerly it must have swarmed with great monsters ; now we find mere pigmies compared with the antecedent

[1] *Creatures of Other Days*, p. 248.

LOSS OF CALCAREOUS ARMOUR IN MAMMALS 209

allied races. If Buffon had known of the gigantic Sloth and Armadillo-like animals, and of the lost Pachydermata, he might have said with greater semblance of truth that the creative force in America had lost its power rather than it had never possessed great vigour.'

What is the reason of this change from the monstrous-sized animals of the past to their modern pigmy descendants? Without carbonate of lime, neither the external nor the internal skeletons can be built up. The same cause—want of sufficient carbonate of lime in the waters, plants, etc.—that caused the rachitis of the dermal skeleton, prevented the internal skeleton from attaining a great size; hence the pigmy animals of *these* days, as compared with the monsters of *those* days. But let us not fancy that it was to the disadvantage of the former. It produced *greater activity* in the carnivora, and *greater fleetness* in the herbivora; and so these have survived, and the ancient unwieldy ones have perished.

Some naturalists seem to think that, as animals became more active, they lost their plate-armour by natural selection, the more active killing out the less active armour-plated animals. But I think that just the reverse may have occurred. Animals became active *because* they began to acquire *rachitic* skins, owing to a lime-famine, this material having been used up mainly in making rocks! The deficiency of lime in the exo-skeleton may have not only brought about more activity in the animal, but deficiency of the same material in the endo-skeleton may have also brought about *smaller* animals; so that the two deficiencies may have gradually contributed to create a *much more active* race of animals than there had been before.

It is not difficult to imagine that, although in some instances the huge dish-cover carapace of an animal like the Glyptodon may have been a good protection, if the animal squatted and hid himself under his huge dish-cover; in other instances, however, this same protection might have become a source of great weakness. Let us suppose that the Glyptodont, as is likely, was sometimes wholly unarmoured on his abdominal surface, like many of the existing Armadillos. Then if his more active enemies had learnt to upset him, and leave him kicking in the air like an overturned Turtle, he would be wholly at their mercy, and they might eat his underside, and scoop him out at leisure, in spite of his formidable dorsal protection. We need not even suppose that each of his enemies required to have great strength, for we know that several Grampuses attack a Whale at the same time and kill it, and that Wolves co-operate in their attacks on large animals.

Such is a speculative sketch of the physical conditions which possibly may have brought about the existing active and smaller races of animals. Briefly, the hugeness of the endo- and exo-skeletons of those ancient extinct monsters may have been due to *excess of lime* in their food.

I have called the absence of lime-stiffening in the exo-skeleton a *rachitic skin*. It is not unlike a rachitic egg-skin, when the hen is unable to obtain lime salts for building up the egg-shell. In the latter case, the deficiency is a great disadvantage, but in the former an elastic and flexible skin, giving freedom to the movements of the muscles and endo-skeleton, would have been a great advantage.

Concurrently with the loss of armour, and with the development of muscular activity, there must have been evolved a larger and more complicated brain. Those huge monsters of ancient times are known to have had tiny brains, in proportion to their prodigious size. The spinal cord no doubt supplemented the small size of the

LOSS OF CALCAREOUS ARMOUR IN MAMMALS

cranial brain, and in some cases, as in the *Stegosaurus*, there is reason to believe that an additional lumbar brain existed, as a part of the spinal cord.

I think I have said enough to indicate by what possible causes the plate-armour of ancient animals may have been got rid of, and therefore shall not follow up this speculation any further.

The upshot has been that the ancestors of the Jaguar and Leopard, from a similar lime-famine, lost their calcareous carapace, which consisted of *bone*-rosettes, and so their descendants have had to be satisfied, with profit to themselves, with rosette *stamps* instead!

I believe that all mammals, including marsupials, descended from *armoured* ancestors, and that a famine of lime by degrees relieved them of the calcareous armour, but left on their skins the stamp of the armour-plates, and, in some cases, that of the carapace as a whole. By slow degrees these also became modified in various ways, and even wholly obliterated. The least altered from the originals, after the armour wholly disappeared, is the Jaguar skin; then the others follow somewhat in this order: Leopards slightly altered; spotted and dappled mammals of all degrees and kinds; more or less altered striped mammals; piebald mammals; mammals with self-coloured bodies, but with ringed tails; and lastly, mammals self-coloured all over.

RELATIONSHIP BETWEEN THE ARMADILLO
THE RHINOCEROS, THE HORSE, THE
GIRAFFE, AND THE ZEBU

'Myriads of types, unready and conservative, have passed out of being; that which they had, but did not improve, has been taken from them, whilst others, by steady improvement of what they had, have had more and more life given to them. But not always by slow, steady increment "during long ages," has the advance been made. Nature does, now and then, make amazing leaps, certain types taking on sudden metamorphosis, and, in the fraction of a lifetime, the low is transformed into the high.'

Professor W. KITCHEN PARKER, *On Mammalian Descent*, p. 93.

PART IX

RELATIONSHIP BETWEEN THE ARMADILLO, THE RHINOCEROS, THE HORSE, THE GIRAFFE, AND THE ZEBU

IN order that we may be able to decipher more fully the hieroglyphics so profusely scattered on the exterior surface of the domestic Horses, we have to study the exterior of the various species of Rhinoceros a little in detail.

Zoologists admit, if I interpret their writings correctly, that the Horse is not only allied to the Rhinoceros, but that very probably the Horse's one digit has descended from the enlarged middle or third digit of some Rhinocero-Tapiroid ancestor.

Then 'the crowns of the lower true molars consist in their simplest structure of two transverse ridges (as in the Tapir), but these ridges may be curved into crescents (as in the Rhinoceros), or complicated by foldings and convolutions (as in the Horse). The transition from the simplest brachydont (or short-toothed), to the most specialised hypsodont dentition (or tall-toothed) is accompanied by a reduction of the number of the digits from four or three to one; that one being the third or middle of the typical series of five.'[1]

If we direct our attention to the hide of the Indian Rhinoceros we seem to feel a sort of conviction that the scapular and pelvic hide-shields of this huge mammal are *homologous* with the bony scapular and pelvic shields of the Armadillos. We seem to feel also

[1] *Manual of Palæontology*, by Nicholson and Lydekker, p. 1353.

that the ears of the Rhinoceros and the Armadillo are singularly alike. Then if we turn to fossils, we find that the Polacanthus has a bony pelvic shield rising in knobs,[1] not unlike those of the hide-shields of the Indian Rhinoceros (Fig. 77). This is what Nicholson and Lydekker say of it (p. 1161): 'In *Polacanthus* of the same (Wealden) beds, we have a remarkable form (of Dinosaur), in which the dermal armour constitutes a complete solid carapace over the

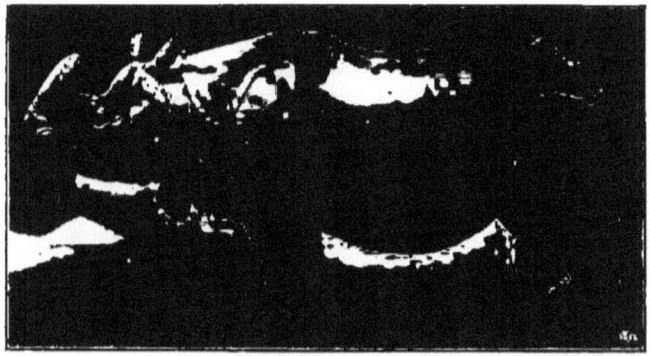

FIG. 77.—Indian Rhinoceros (*R. Unicòrnis*), from a photograph, F. G. O. S. 20046.

whole of the dorsal aspect of the lumbar region, some of the component scutes being tuberculated, and others ridged ; while there was also a number of detached flattened spines somewhat like those of Hyæosaurus, which probably formed a line in the dorsal region. This peculiar type of carapace forcibly recalls that of the Glyptodont edentates.'

In order to emphasise the similarity of the Rhinoceros and the

[1] See Fig. 63 (a).

RELATIONSHIP OF ARMADILLO TO OTHER MAMMALS 217

Armadillo, I would here mention two other interesting points. The Giant Armadillo (*Prionodon Maximus*) has its hind feet *ungulate*: its hoofs are almost exactly like those of the Malayan Tapir ; and in some Rhinoceroses the incisor teeth are wholly wanting, and that part of the jaw is contracted, not unlike that of the Glyptodon.

Then there is a type of Rhinoceros which brings us still more closely to the Armadillo type. This is what Mr. W. T. Blanford[1] says of the *Rhinoceros Sondaicus*, the smaller one-horned Rhinoceros which has been observed at great elevations in Burmah: 'Skin naked, or nearly so, not tubercular, the epidermis divided by cracks into small polygonal, sub-equal scale-like discs throughout the body and limbs. Surface of the body divided into shields by folds, as in *R. Unicornis*, but the fold in front of the shoulders is continuous across the back, like that behind the shoulders and that in front of the thighs.'

That is to say, unlike the Indian Rhinoceros, the divisions of the plated hide of this Javan Rhinoceros go right over the back exactly like the divisions of the carapace of an Armadillo. So that between the type of the Glyptodon and the type of the existing Armadillo there must have been some intermediate type approaching that of the *R. Sondaicus*, in which the armour still consisted of *bone-rosettes*, somewhat like those of the Glyptodon, but the whole carapace admitted of a freer movement, because it was divided into three or four separate sections attached to each other by simple unarmoured skin, acting as leather hinges, and the *Rhinoceros Sondaicus* may be the descendant of this intermediate type of Armadillo.

Now in the Natural History Museum there is a small Rhinoceros without a ticket (at least the ticket is not visible) about the size of

[1] *Fauna of British India*, 'Mammalia,' p. 474.

a Newfoundland Dog. The specimen is hidden by the larger ones, and jammed in among them, so that it is not easy to examine it thoroughly.[1] It has no horns, or perhaps only a rudimentary one on its nose. It may be a young one of *R. Sondaicus* (?) Of whatever species it may be, it presents quite a revelation as to the points I am discussing. It has the scapular and pelvic shields somewhat like those of the Indian Rhinoceros (*R. Unicornis*), but the whole surface is covered with small plates, which vary as to their component platelets.

In Fig. 78 I have given two variants taken from the posterior aspect of the left haunch of this Rhinoceros. Its plates appear to be of some *bony* substance.

FIG. 78.—(*a*) and (*b*) Bony plates from haunch of small Rhinoceros, Natural History Museum (*R. Sondaicus?*); (*c*) bony plate of *Tolypentes tricincta* (an Armadillo).

This is not all, for in one or two places the plates have fallen off, showing that they are something quite distinct from the dermal tissue proper; and what may perhaps be of some importance, the detachment of the plates has left an *impression* of their component platelets on the subjacent skin.

Some of the plates of this little Rhinoceros, like (*a*), do not differ materially, although they do in outline, from those of one of the Armadillos (*Tolypentes tricincta*), shown in (*c*). Then (*b*) in its component platelets is not unlike a Jaguar rosette.

[1] Its position has been recently altered.

RELATIONSHIP OF ARMADILLO TO OTHER MAMMALS 219

The curious part is that the tail of this Rhinoceros is also covered with similar plates *disposed ring-fashion like those of the tail of an Armadillo.*

Here then we have a Rhinoceros encased in armour, almost exactly in essence like that of an Armadillo.

A glance at the other species of Rhinoceros in the Natural History Museum teaches us something more in our endeavour to discover further evidence of the relationship of these mammals and the Horse. For *R. Unicornis* (India) has the skin covered with *hide* tubercles only, without any deposit of bone-plates, admitting of a freer movement. *R. Sumatrensis* (two-horned) is the smallest and the most hairy, the skin being only rough and granular with the folds of skin less marked. It is evidently a type which has passed from the plated to the smooth and hairy mammal. This also is met with at a high elevation (4000 feet). There are variations of the Sumatran Rhinoceros from Chittagong and Malacca, which have a still smoother skin and longer hair. We come ultimately to the African species, which have neither plates, nor folds, nor hair of consequence, and their skin is nearly as smooth as that of a Porpoise, making allowance for the difference of the medium in which these two kinds of animals live.

Then in Fig. 79 I have given what appears to me a variation of *Rhinoceros Sondaicus*. It is not only covered with small plates, from head to foot, but it has a curious, and, may be, instructive variation of plating on its flank. About this I have something to say in another place. In the meantime, in Fig. 80 I give four groups of plates, picked out from what I suppose to be a *R. Sondaicus* (Fig. 79). These groups are not unlike the rosettes of the Armadillo of Fig. 78 (*c*).

It will be seen that the Indian Rhinoceros of Fig. 77, although it has lost its bone-rosettes, has retained hide-knobs, disposed

220   STUDIES IN THE EVOLUTION OF ANIMALS

as shown in Fig. 81. (*d*) is a group of spots from the shoulder of a Horse, given here for comparison.

FIG. 79.—*Rhinoceros Sondaicus* (?), from a photograph by Messrs. Dixon and Son.

FIG. 80.—Plates picked out of the Rhinoceros of Fig. 79. (*a*) and (*b*) from middle of flank; (*c*) from hind margin of scapular-shield; (*d*) from hind margin of neck-shield.

The hide-knobs of the Indian Rhinoceros shown in Fig. 81 are represented as seen from a certain distance on the animals of the Zoological Gardens.

RELATIONSHIP OF ARMADILLO TO OTHER MAMMALS 221

Some of the stuffed Rhinoceroses of the Natural History Museum are so placed that it is impossible to get near enough to them to examine their hide in detail. In the Edinburgh Museum of Science and Art, however, I had a rare opportunity of

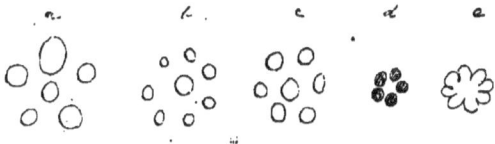

FIG. 81.—(a) and (b) are from the shoulder and hip of an Indian Rhinoceros; (c) from the hip of another individual of the same species (Zoological Gardens); (d) is a group of black spots seen on the shoulder of a Horse; (e) hide-plate from Hairy Rhinoceros of Chittagong (Zoological Gardens).

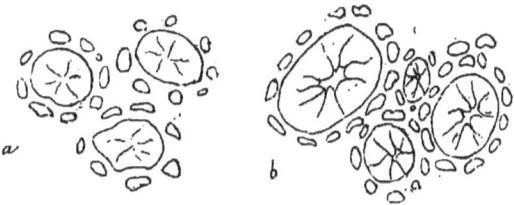

FIG. 82.—(a) From the lower and back part of the scapular shield; (b) from the back part of the pelvic shield, of an Indian Rhinoceros (Edinburgh Museum of Science and Art) —reduced.

minutely examining the hide-shields of an Indian Rhinoceros (*R. Unicornis*).

From Fig. 82 it will be seen that the hide-knobs of this animal resolve themselves into patterns almost identical with the armour bone-plates of the Sturgeon and the Crocodile shown in Fig. 68 (*b*) and (*c*), and of the Glyptodon shown in Fig. 61 (*g*).

The central portion of these rosettes is a pyramidal hide-knob, composed of smaller sections, and surrounded by the usual ring of small tubercles. Compare these with the Horse rosettes of Fig. 56 (*a*).

The similarity of pattern between the hide-knobs of this Rhinoceros and the bone-rosettes of the Glyptodonts and Armadillos is astonishing, and leaves little doubt in one's mind that the Indian Rhinoceros, with its scapular and pelvic shields and its hide-armour, comes down to us from some Armadilloid ancestor.

In the Horse, every trace of plate and tubercle has disappeared, but they have left instead their indelible *pigment-pictures* in some varieties of the dappled Horse, as shown in Fig. 56 (*a*), (*b*) and (*c*).

We have now to examine a very interesting feature of the *R. Sondaicus* given in Fig. 79. Its flanks look as if they were cracked into larger plates than those of the rest of the body. They may be in fact *fusions* of a number of the smaller plates—for although the plates of this Rhinoceros are of a bone-like nature, they may be quite capable of fusion as shown in the hide-plate of the Chittagong Rhinoceros (Fig. 81 (*e*)).

It would at first seem improbable that stiff bone-rosettes are capable of fusing into larger plates; but we have already seen an approach to fusion of bone-rosettes in the armour of both the Glyptodon and the Leathery Turtle (Figs. 67 (*a*) and 68 (*a*)).

This cracked appearance of the hide of *R. Sondaicus* of Fig. 79 forcibly recalls, it seems to me, that flank reticulation which is so common a feature in the dappled Horse (see Figs. 34 and 38). And in both animals it may mean the same thing, viz., in the Rhinoceros the fusion of one or more *bone*-rosettes, and in the Horse the fusion of one or more *picture*-rosettes. If this be conceded, we come to the same conclusion which has been reached frequently before, viz., that both the Rhinoceros and the Horse have descended from some Glyptodontoid ancestor.

Of course Glyptodontoid and Armadilloid animals vary in their bone-plating, as much as Rhinoceroses vary in their hide-plating; so do Horses and other mammals vary still more extensively in their picture-plating.

In some species of Rhinoceros the shields and plates have entirely disappeared. Having got rid of the plates, some of the species got rid also of the hide-shields. Natural selection was evidently rendering their skin more pliable, in order that their movements might be freer. In other words, their movements became freer as they were relieved of their thick hide-plates and shields. Not improbably the African species, as a compensation for their loss of armour, may have obtained those formidable horns which we see in the Natural History Museum.

In the dappled Horse the pigment arrangements have been so modified in most cases, as hardly to maintain any resemblance to the typical rosetting of the Jaguar. In only a few cases have I succeeded in tracing true rosettes in the Horse, and these I have shown in Fig. 56. Usually it is hopeless to trace any minute rosette resemblance in the Horse dappling. But by observing an infinite number of all colours, and putting two and two together, one may succeed in evolving an ideal of what some remote ancestor of the Horse may have looked like. Anyhow, the Zebras are clear evidence that, whatever the external disposition of pigments may originally have been, the resulting modification has been almost identical with that of the Tiger. In this, and much more clearly in other Cats, it is easy to prove to one's-self that stripes and bands originated from rosettes like those of the Jaguar and Leopard.

Then in striped animals themselves we seem to witness modifications undergoing before our eyes, so to speak. In Grevy's Zebra the bands are very numerous and narrow, some of them coalescing into one band; while in Burchell's Zebra every alternate band has become fainter, and may be disappearing altogether.

I have so far endeavoured to trace the relationship of the Armadillo, the Rhinoceros, and the Horse, simply from characters which they retain on their hides. This relationship I find can be also traced to the Zebu of Fig. 58, and to the Giraffe of Fig. 57. In another place I have shown that probably there is a more intimate relationship between the Horse and the ruminants than there is between the Horse and the Rhinoceros. Of course their skeleton and general anatomy sufficiently indicate a certain fundamental relationship, but what I am aiming at here is to trace a closer relationship between certain groups of mammals by means of the hieroglyphics exhibited on their skin. With these external and more transient characters anatomy does not concern itself.

If the reader will take the trouble to compare certain Horse rosettes shown in Fig. 43, certain Giraffe rosettes shown in Fig. 83 (*a*) and (*b*), and the rosettes of the Zebu of Fig. 83 (*c*) and (*d*), with certain Jaguar rosettes of Fig. 59, and then again compare certain bone-rosettes of Armadillos of Fig. 63 (*c*) with certain hide-rosettes of the Rhinoceros of Fig. 80, he can hardly fail to be convinced that they *all*—bone-rosettes, hide-rosettes, and pigment-rosettes—had *one and the same origin.* Of course the rosettes of the Horse, and more especially those of the Ox, have undergone much alteration, not only through remoteness of descent, but also by prolonged domestication, and human selection, so that piebaldness and total obliteration of all traces of rosetting are very frequent features in these animals.

From what evidence I have been able to adduce, I think we are justified in again drawing the broad conclusion that, in mammals at least, all spotted, striped, ringtailed, and piebald animals have descended from armour-plated ancestors. And as we know that

RELATIONSHIP OF ARMADILLO TO OTHER MAMMALS 225

both among wild and domesticated mammals a number of self-coloured varieties are indubitably related to either spotted or striped congeners, we reach again a still broader conclusion that very probably *all* mammals, *including the marsupials*, have descended from armour-plated ancestors.

Having once surmised that the reticulations we see on the

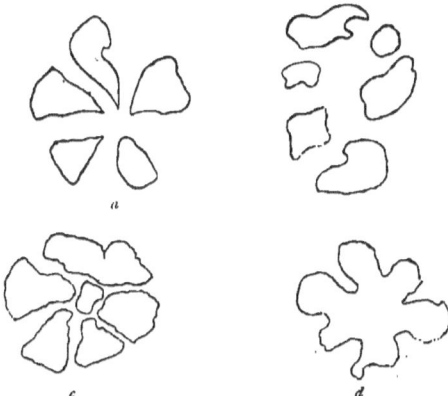

FIG. 83.—(*a*) Rosette from lower part of right haunch of Giraffe; (*b*) rosette from right shoulder of the same (Natural History Museum); (*c*) rosette from upper part of fore-leg of Zebu; (*d*) rosette from shoulder of the same (see Fig. 58).

dappled Horse's flank may be *imprints* of the commissures between the plates which we see on the flank of *Rhinoceros Sondaicus* of Fig. 79, something further is suggested.

In Fig. 40 I have shown that the commissure-pictures *coincide* with the network of superficial veins on the Horse's flank. Not improbably these superficial veins are the ancestral *commissure-veins*; that is, a network of veins of a certain size located in the

P

clefts of the plate-commissures, into which the blood supplied to the plates found its way much as the intra-lobular veins of the liver receive the blood from the veinlets of the lobes. Whether this be so or not may perhaps one day be ascertained by dissection of the skin of the Rhinoceros, and also of the skin of the Armadillo. It might thus be easily ascertained whether the commissures of the plates have any corresponding veins.

That the superficial veins of the Horse, or their nervous apparatus, play some part in the distribution of pigments seems likely. The ungulate character of the Horse's hands and feet would naturally lead one to surmise a close relationship between this animal and the Rhinoceros, but we should not forget that the Great Armadillo on its hind feet has *hoofs*, and not claws.[1] Both the Horse and the Rhinoceros furnish indications of descent from carapaced ancestors, but the question is whether the Horse came to be what it is through the ruminant branch, or through some Rhinocerotoid branch of ancestors. In another place I have shown that there is reason to believe that the Horse and the ruminants are closely allied. From features that I have indicated in the *R. Sondaicus*, especially in his astonishing tail, encased in *rings of bone-plates*, there can be no reasonable doubt that *he*, at all events, descended from armour-plated ancestors.

The degradation from bone-armour to picture-armour might at first sight appear a great disadvantage for subsequent races, but in reality it has resulted in giving natural selection a totally different direction. It has resulted in the evolution of fleetness in the Horse, Antelopes, and Deer, and in the astonishing springs of the carnivora that feed on them. In other words, it has resulted in the development of great muscular and brain power—with a suitable

[1] See also p. 132, *Mammalian Descent*, by Professor Parker, about 'ungulate Lemurs.'

RELATIONSHIP OF ARMADILLO TO OTHER MAMMALS 227

framework of bone—the lines on which the animal world was *thereafter to be ruled*.

It would be idle to suppose that the bony-plates of the Armadillo, the hide-plates of the Rhinoceros, and the picture-plates of the Horse, are all so like each other by *mere accident*, any more than we can consider that the seven cervical vertebræ of these animals came to them all by accident. One can fully understand that the external features of an animal are subject to much greater variations than the internal skeleton; nevertheless, in spite of all the modifications in the skin, which the animal unconsciously brought about in order to adapt itself to its surroundings, enough is left in the skin features of many species to assure us that all these animals have had a common origin, and that they descended from an *armour-clad ancestry*. The evidence of the pictures on the skin of the Jaguar, as I have shown, is too clear to admit of any other interpretation than that of descent from armour-plating.

I have said enough to prove that the markings on the skins of mammals are not necessarily the result of natural selection, working gradually on some incipient variation through long periods of time; but more probably the result of an *inherited nerve-centre habit*. Then, after the total disappearance of calcareous matter from the skin, this habit continued its action on the pigments of the skin, and was modified subsequently in a hundred ways. What natural selection probably *did do* was, as I said, to *maintain* this inherited skin feature, wherever it was *advantageous* to the animal, and to obliterate it wherever it was not.

From all this wearisome discussion we learn one great lesson, and it is this:—that habits, or memories, of the nervous system, when established for long ages, are got rid of *very slowly*; and that it requires more long ages for *new* ones to become established in their places.

EXPLANATION OF THE CALLOSITIES ON THE LEGS OF EQUINE ANIMALS AND OTHERS

'We find *heredity*, or adherence to a general type derived from ancestors, opposed by special modifications of or deviations from that type, and the latter generally getting the victory, although in the numerous rudimentary structures that remain there is significant evidence of ancestral conditions long passed away.'

The Horse, by Sir W. FLOWER, p. 3.

PART X

EXPLANATION OF THE CALLOSITIES ON THE LEGS OF EQUINE ANIMALS AND OTHERS

ONE of the features which distinguish the Horse from the Ass is the presence of callosities on the inner aspect of *all four* legs in the Horse, while in the Ass and Zebra these callosities are present only on the *fore-legs*. In the Horse, on the fore-legs, they are above the wrists (commonly called knees); and on the hind-legs, they are below the heels (commonly called hocks).

One day, at the sea-side, I was examining some little Donkeys which had stripes on their legs. I asked the Donkey-boy, 'What are those things on the inside of the Donkey's fore-legs?' He said, 'I don't know, I will go and ask'; he added, 'but I can tell you what that mark on the Donkey's back is.' 'What is it?' I said. He replied, 'When Jesus Christ went to Jerusalem he rode a Donkey, and after that it had the sign of the cross on its back.' I suppose he inferred that his little Donkey was a descendant of *that*. The event of the entrance into Jerusalem, be it remembered, must have been a good bit before the Crucifixion!

A well-dressed man who was standing by and heard our conversation said—'Every gentleman knows what those things are on the Donkey's fore-legs. They are corns.' 'But,' said I, 'they are very different from our corns.' 'Yes,' he observed, 'the corns of Horses and Donkeys are different.'

Stable-men call them corns. Professor Flower calls them warts or callosities, or 'chestnuts,' and 'mallenders,' and 'sallenders,' as some old books designate them.

He says—'There the skin is peculiarly modified from its usual structure.' No hair grows on these patches; 'the papillæ of the derm or true skin are much enlarged, and covered with an abundant and thick epidermis, which becomes dry and horny, and sometimes (in the Horse) accumulates in considerable quantity on the surface, occasionally even making a horn-like projection.'

Professor Flower[1] adds—'The signification and utility of these structures are complete puzzles;' and on p. 173, 'They obviously belong to a numerous class of special modifications of particular parts of the cutaneous surface which occur in very many animals, the use of which is in most cases remarkably obscure. Bare spots, thickened patches or callosities, and tufts of elongated hair, often associated with groups of peculiar glands, are very common on various parts of the body, but especially the limbs, of many ungulates, and to this category the "chestnuts" of the Horse undoubtedly belong.'

Then, in a note to the same page, he says—'Dr. J. E. Gray divided the pigmy Chevrotains (ruminants allied to Pigs) into two groups—*Meminna*, 'with a naked prominence on the outer side of metatarsus, rather below the hock,' and *Tragulus*, 'with hinder edge of the metatarsus naked and callous.'[2]

Professor Flower concludes this part of his book thus: 'If they (the corns) teach us nothing else, they afford a valuable lesson as to our ignorance, for if we cannot guess at the meaning or use of a structure so conspicuous to observation, and in animals whose

---

[1] *The Horse*, p. 172.
[2] At the end of Part VI. I made a suggestion which may give a different meaning to the callosity of Tragulus.

CALLOSITIES OF EQUINE ANIMALS AND OTHERS 233

mode of life more than any other we have had the fullest opportunity of becoming intimately acquainted with, how can we be expected to account off-hand for the endless strange variations of form and structure of animals of whose habits and methods of existence we know absolutely nothing?'

I had pondered over the callosities of the Horse, Ass, and

FIG. 84.—(*a*) Callosity on fore-leg of Tibetan wild Ass or 'Kiang' (*Equus hemionus*); (*b*) callosity on inner aspect of hind-leg of the 'Alpaca' (*Lama pacos*); (*c*) tuft of dark hair on light ground of leg of the 'Vicugna' (*Lama vicugna*); all three from Natural History Museum.

Zebra a good bit before I read the foregoing in Sir W. Flower's book on *The Horse*.

Fig. 84 (*a*) shows the callosity on the fore-leg of the Tibetan wild Ass; and Fig. 53 shows it on the fore-leg of the Zebra.

I have never met with an Ass which had callosities on its hind-legs. In the stables of the Great Northern Railway Company I have seen a Horse which had *no callosities on the hind-legs*; and

234 STUDIES IN THE EVOLUTION OF ANIMALS

another which had them as small as peas on those limbs ; and the foreman of the stables of the London Road Car Co. told me he has seen them as small as the nail of his little finger. It is therefore not improbable that the ancestors of the Ass and Zebra may have had callosities also on the hind-legs, but from some

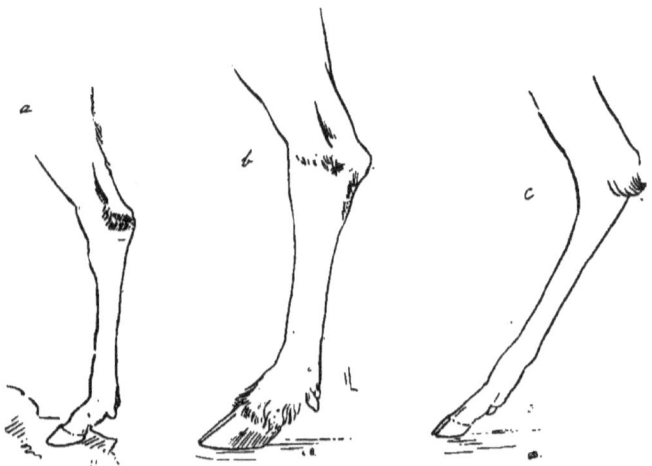

FIG. 85.—(a) Black tuft of hair on fawn-grey—hind-leg of Peruvian Roebuck (*Coriacus antisiensis*) ; (b) dark brown tuft on a dirty white ground—hind leg of Elk ; (c) grey tuft on fawn-coloured leg—hock of Virginian Deer (*Coriacus Virginianus*)—from Natural History Museum.

cause they may have dwindled off and disappeared so completely from those species that they are not reverted to. As far as I am aware, if the Ass and Zebra ever had them on their hind-legs, no trace of them is to be seen there now.[1]

[1] It is curious that *Equus Grevyi* of Fig. 52 does not show a callosity on the fore-leg as in Fig. 53. Grevy's Zebra which I saw at Tring also has its fore-legs *without* them.

CALLOSITIES OF EQUINE ANIMALS AND OTHERS 235

In the domestic Horse they vary in shape. In the fore-legs they are either pointed up and down, or, as in Fig. 84 (*a*), pear-shaped; and in the hind-legs they are often like slits in the skin. I have not seen a second case of a Horse without them on the hind-legs, but perhaps a more extensive search might discover some others.

Now, are there any other animals with similar callosities on their legs? Yes; Professor Flower has mentioned the *Meminna* and *Tragulus*,[1] two Deer-like or Swine-like ruminants. In the Natural History Museum the Huanaco (*Lama huanacus*) has similar callosities on both the inner and outer aspects of the hind-leg below the hock. The Llama (*Lama glama*) has them also on both sides of the hind-legs. The Alpaca (*Lama pacos*) in some cases seems to have them only on the inner aspect of the hind-legs.[2]

Fig. 84 (*b*) gives an outline of the hind-leg of the Alpaca, showing the callosity; and Fig. 86 gives a full picture of a living Llama, with the callosities, as slits, on both aspects of the hind-legs.

It is interesting to note that in the Vicugna (Fig. 84 (*c*)), another Cameloid animal of South America, instead of a callosity, there is, in this case at all events, in the same region, a distinct tuft of elongated dark hair on a light ground.

Then again in the Natural History Museum there are various other ruminants which have, as I think, very significant tufts of hair on the inner aspects of their hind-legs which are of a different colour from the general ground colour.

The Elk, the Virginian Deer, and the Peruvian Roebuck—all American ruminants—have these suspicious tufts of hair on their

[1] The callosity on the metatarsus of Tragulus may mean something else (see end of Part VI.).
[2] At Tring, the Vicugna has callosities, like slits, on both aspects of hind-legs, below the hock; and in the Zoological Gardens both the Huanaco and the Llama have callosities in the corresponding places.

236   STUDIES IN THE EVOLUTION OF ANIMALS

hocks. In all cases they are of a different colour from the surrounding hair. Fig. 85 gives an outline of their legs. But the most interesting of all is the male Reindeer of the Natural History

FIG. 86.—The Llama, from a photograph, F. G. O. S., 20040.

Museum. On the inner aspect of its hock it has a large whitish patch of hair on a dark ground.

There can be little doubt, I think, that these singular tufts of elongated or modified hair are, as Sir W. Flower suggests, vestiges

CALLOSITIES OF EQUINE ANIMALS AND OTHERS 237

of *once glandular* surfaces of the skin, which have ceased to be of any use to the animals that possessed them, and have now ceased to have selection value. In some, as in the equine family, they have dwindled into mere callosities with thickened epidermis, while in others they have become covered with hair of a different tint and length from the surrounding hair. In the Camels of South America, with possibly certain exceptions, they appear to be in the same condition as those of the Zebra and Wild Ass, only much reduced in size.

It would seem safe to look upon all these callous surfaces as vestiges of skin glands, similar to those in front of the eye of certain Deer, like that shown in Fig. 87.

FIG. 87.—Gland in front of the eye of Peruvian Roebuck (*Coriacus* (*Furcifer*) *antisiensis*).

Sheep have glands between the hoofs of all four legs; and goats have them only between the hoofs of their fore-legs, and sometimes not even there.[1]

Now we come to another and more obscure point of our research. What could have been the use of skin glands on the legs of the common ancestors of all these different mammals—Horses, Asses, Zebras, Deer, and Camels?

Well, in the ancestral history of these animals, skin glands on the legs, which secrete an odoriferous substance, may have been of

[1] *Royal Natural History*, vol. ii. p. 234.

the utmost service for purposes of recognition and discovery, by leaving traces on the high grass through which they may have passed, or in which they may have hidden themselves. Mr. F. C. Selous mentions that on being attacked by natives one night, he and his men escaped into the jungle grass, which was *seven feet high*.[1] In the midst of tall grass, neither keen sight nor keen hearing would be of much avail, but by leaving a scent on the grass which rubbed against the leg glands, as the animal passed through, its companions might succeed in rejoining it. When hunted by a carnivorous animal, it would have been an advantage for a herd to scatter, and hide themselves among grass. If they did so, it would have been an advantage to be able to find each other again by leaving scent-traces on the grass. Sir W. Flower says—' Some Deer and Antelopes have a suborbital gland secreting a peculiar oily odorous substance.' When the animal is feeding drops fall on the herbage and indicate to others the whereabouts of animals of the same species. It is not improbable that the leg glands had some similar use.

Just imagine how useful such a trace would have been to the dam in finding its young one, and to the young one in finding its dam, when they may have wandered away from each other in the midst of high grass.

Think how useful they would have been in the rutting season among high grass to the male in finding the female and *vice versâ*. As I said, such glands were very probably of the same nature as those in front of the eye of certain Deer.

The reader might perhaps ask—How is it that *all* the ruminants have not traces of leg glands? Well, the same question might be put regarding the suborbital gland, for not all Antelopes and Deer have it. We don't know sufficiently the early history

[1] In South America miles and miles of plains are covered with the tall Pampas grass.

CALLOSITIES OF EQUINE ANIMALS AND OTHERS 239

of each species from the time it branched off from the common stock. For all we know, ruminants may have had leg glands, but some compensating condition may have rendered that feature of little importance in some of them, and a cessation of selection in that direction occurred, by which the leg glands were eventually totally suppressed, while in others traces of them have been left in the form of patches of hair of different colour and length from the surrounding hair.

As the Horse has them on all four legs we may assume that Zebras[1] and Asses had them also on their hind-legs, although now they have no traces of them there. It is not only callosities, but other features may become often lost by *suppression*. It is one of the factors in the origin of species.

I have never found anything recorded by hunters of Zebras on these leg glands. In these animals the callosities are very large, and something, one would think, might be discovered about their real nature and structure if proper investigation were made when the Zebra was fresh after being shot.

It would be very interesting also to ascertain whether in the Horses which have become again semi-wild in South America and Australia, these skin glands have ever re-acquired any of their ancient activity. Not impossibly the ancestral possessors of these leg glands may have been driven by their enemies from regions of long grass to more open regions, and rocky places with shorter grass; or the long grass of certain regions may have been devoured by the immense multitudes of these animals which are recorded by hunters; and so the leg glands may have become less useful, and by degrees lost their selection value. Anyhow, it seems more than probable that those singular tufts of hair were in some ancestral form active skin-glands of some use to

[1] Grevy's Zebra seems to have lost them on both hind- and fore-legs.

the animals that possessed them. And the callosities of the equine family are probably vestiges of similar glands. The equine and the ruminant families of mammals would appear to have branched off from the same stock. In another place I have given reasons for considering these two sets of animals as closely allied.

In my opinion there cannot be much doubt that Sir W. Flower's suspicion, as to the nature of the callosities of the Horse, is the right one. They 'belong undoubtedly,' he says, 'to the category of glandular surfaces.' They appear to me to be vestiges of skin-glands which once served a useful purpose in past ages, and were not unlike the suborbital glands of some kinds of Deer.

Innumerable changes may have naturally occurred on the face of the countries their ancestors inhabited; and the advent of man, of whose antiquity there can be little doubt, has made still further changes which may have rendered those glands useless, if not hurtful should they become injured.

Mr. Louis Robinson[1] has stated that whoever discovers the meaning of the Horse callosities 'will become famous among naturalists all the world over.' Their meaning has already been suggested by one who is 'famous all the world over'—viz. Professor Sir William Flower.

[1] *North American Review*, April 1894, p. 483.

THE ONE BIG DIGIT OF THE HORSE

Q

'Numerous "Lemurs" with marked ungulate characters are being discovered in the older Tertiaries of the United States and elsewhere. . . . I am not aware that there is any means of deciding whether a given fossil skeleton, with skull, teeth, and limbs almost complete, ought to be ranged with the Lemurs, the insectivora, the carnivora, or the ungulates.'

Professor HUXLEY—*Proceedings of the Zoological Society*, 1880, p. 65.

PART XI

THE ONE BIG DIGIT OF THE HORSE

PROFESSORS of Biology and Palæontology have been teaching that the Horse has only *one* enlarged digit in its four limbs, with vestiges of two other digits hidden under the skin (the splint

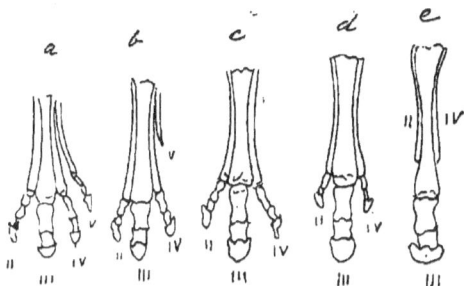

FIG. 88.—(*a*) Foot of an extinct form of Horse-like animal, *Pachynolophus* (Eocene); (*b*) of *Anchitherium* (Early Miocene); (*c*) of *Anchitherium* (Late Miocene); (*d*) of *Hipparion* (Pliocene); (*e*) of *Equus* (Pleistocene), and also of the existing Horse;—from *Mammals*, by Flower and Lydekker, p. 377.

bones); and that this big digit owes its origin to one of the digits (the iii.) in the hand and foot of an extinct animal, repeated with modification in several other descendants, also extinct, until we reach that of the Horse—fossil and existing—as shown in Fig. 88.

This enlarged digit is traceable in some other existing animals of the same line of descent, or branches thereof, such as the Rhinoceros, the Tapir, etc.

In the Horse it is stated that all other digits have disappeared, leaving only what are called the 'splint bones'—insignificant remnants, or vestiges, which show that originally two other digits were there. The big one only has remained, becoming bigger and bigger by absorbing to itself the function of all the normal five digits. Indeed, in the modern Shire stallion, the hoof is not unlike a round dish-cover.

In the Natural History Museum [1] there is arranged a very pretty gradation of specimens, commencing from an ancient extinct type, which has one big digit (the iii.) in the middle of four smaller ones, and going up to the one solitary digit of the Horse.

It has become a sort of dogma that the hand and foot of the Horse are different from the hand and foot of the Ox; the former being, it is supposed, *one* enlarged digit, and the latter *two* distinct digits; and further, that these two animals belong to two different groups, the Horse belonging to the uneven-toed group (*Perissodactyls*), and the Pig, Ox, etc., to the even-toed group (*Artiodactyls*).

As so many professors of note have embraced and propounded this theory, it might seem a bold sort of presumption on my part to re-open the question, which has become the basis of an established conviction among biologists. But facts, they say, are stubborn things. I shall place before the reader certain facts, which are my reason for re-opening this interesting question.

In discussing a subject like this it is well to begin at the beginning. In the Natural History Museum, gallery of fossils, case 9 (c. d.), there is a cast of an extinct animal—*Phœnacodus primœvus* (Eocene)—not larger than a biggish Dog, with five digits

[1] Fossil Gallery, case 9.

THE ONE BIG DIGIT OF THE HORSE 245

in its hand and foot. Each digit ends in what is supposed to have been a small hoof, so that this little extinct animal would appear to have been a five-hoofed mammal. It has the middle digit (the iii.) of each limb enlarged. So there is evidence in support of the possibility that this enlarged middle digit of *Phœnacodus*, or of some animal like it, may have persisted in the innumerable modifications which must have occurred, and have become the ancestor of the solitary digit of the modern Horse. Indeed, the evidence of the gradations from the former to the latter by gradual loss of all the digits, excepting the middle one (the iii.), seems complete and satisfactory.

In Fig. 88, it will be seen that *Pachynolophus* had already lost one of its Phœnacodal digits, viz., the big toe. Some weakness of innervation and diminution of circulation in a part will lead to disuse, and consequent dwindling, and eventual suppression.

All the specimens in the Natural History Museum are being so beautifully arranged that all who can read may learn. One thing is wanted there—viz., a lecturing theatre, and some endowment left by some rich person for lectures, to teach the people the great philosophy of evolution.

There are known several other allied extinct animals with a hand and foot, probably derivable from some five-digited animal like *Phœnacodus*, such as *Hyracotherium venticolum*, *Palæotherium*, etc.

Mr. Hutchinson[1] makes the following suggestion: 'One cannot help sometimes wondering whether to some extent the *will* of an animal may not be an important factor in evolution, although it is the fashion to ignore it, and to attribute organic changes to natural selection, or the " survival of the fittest." Mind has a powerful influence over matter,[2] and can we not conceive

[1] *Creatures of Other Days*, p. 214.   [2] Yet the two are *inseparable*.

that the earliest ancestors of the Horse, such as *Phænacodus*, finding (perhaps for the first time in the world's history) boundless grassy plains before them, were impelled by a strong desire to run? The exercise itself must have been enjoyable and invigorating; besides, the more ground they could cover in a day the greater choice of pasture they would find, while at the same time increased swiftness of foot meant greater safety from flesh-eating enemies.'

The 'will,' being dependent on some special centre in the brain, cannot possibly be excluded from being a factor, or part of a factor, in natural selection. Probably there is no part of the organism which does not contribute its quota towards the *co-operative factorship* in natural selection. As to the running being 'enjoyable and invigorating,' one has only to observe a young Foal in a paddock, or the young Eland in its paddock at the Zoological Gardens, to be convinced of the enjoyability of galloping round and round the paddock, as if imaginary devils were running after it. These young animals seem to do it with no other object than that of 'enjoying a gallop'—a sort of exuberant *joie de vivre*, no doubt impulsed by plenty of nourishing mother's milk.

The reader who may not have paid much attention to the interesting doctrine of evolution should not suppose that, because the dray Horse is of elephantine dimensions, it could not have descended from an animal as small as a big Dog; for there have been pigmy Elephants and giant Elephants, pigmy Deer and giant Deer, pigmy Lizards and giant Lizards, pigmy Men and giant Men, and so forth. Large size would appear to be rather a matter of luxuriant feeding which contains all the ingredients of bone, muscle, nerve, etc., and of ease in obtaining it, while small size is the reverse, the structure of the two—large and small—remaining, bone for bone, muscle for muscle, identical, and therefore evidence of derivation from a common stock. Of course a giant

THE ONE BIG DIGIT OF THE HORSE 247

Elephant could not pair with a diminutive one, and thus the two breeds would remain distinct.

Mr. J. I. Lupton[1] is of opinion that 'The land upon which every species of animal is bred exerts a powerful influence upon it for good or evil. We find that a sterile soil will stunt the growth of animal life, whereas a rich one will promote vigour, size, and stoutness.'

Although abundance of food, or the reverse, may have a great deal to do with giant and dwarf forms, still the abnormal size may have been initiated as a montrosity *suddenly*.[2]

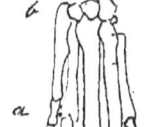

Then the Pachynolophus form of foot would seem derivable from a Phœnacodus form of foot. The latter has five digits, with the third enlarged, while the former has only four, with the corresponding, or homologous one, also enlarged, while its big toe is suppressed.

FIG. 89.—Abnormal hand of a Pig, (*a*) being the thumb, and (*b*) the trapezium, one of the wrist bones *fused* with the palm bone of the thumb; from Fleming's translation of Chauveau's *Anat. Comp. des Animaux domestiques*, p. 122.

It should be noted that, as an anomaly, the hand of the Pig sometimes comes out with five digits as shown in Fig. 89. The curious part of this anomalous Pig's hand is that one of the carpal bones,—the trapezium (*b*)—is completely fused with the metacarpal bone of the thumb.

From this anomalous hand it would appear that the sub-division of the ungulates or hoofed mammals into uneven-toed and even-toed is rather arbitrary, for here is an even-toed Pig giving birth to an uneven-toed child. Of course both sub-divisions originated from the same stock; but two Greek names to indicate a condition

---

[1] 'Pedigrees of British and American Horses' (*Nineteenth Century*, June 1894, p. 926).
[2] See 'Nanisme et Geantisme,' Guinard, *Précis de Tératologie*.

which is not permanent might be dispensed with as advantageously as the two suppressed digits of the Horse!

The reader might say that from the series of limbs shown in Fig. 88 it is clear that the one digit of the Horse is derived from the big middle digit of some animal like *Phænocodus*. Suppression of parts is seen to be such a common phenomenon in the process of evolution throughout the animal kingdom, nay, throughout the whole of what we call living nature, that it is the most rational conclusion to come to. Allied animals had five digits, therefore those that have less have lost some by suppression.

With reference to the evolution of the Horse, as evidenced by the feet of the fossil forms of the Horse tribe, Dr. Alfred Russel Wallace[1] writes : 'Well may Professor Huxley say that this is demonstrative evidence of evolution ; the doctrine resting upon exactly as secure a foundation as the Copernican theory of the motions of the heavenly bodies at the time of its promulgation. Both have the same basis—the coincidence of the observed facts with the theoretical requirements.'

There can be no doubt about the evolution of the Horse from some such ancient form as that of the *Eohippus*. The question in this discussion is restricted to the origin of the big digit in the whole series, that is, whether it was originally *one* digit, or a fusion of *two*.

I might perhaps say that I do not dispute that the origin of the existing Horse's digit is as stated, but I would now ask—Are we so 'cocksure' that *Phænacodus* is to be credited with only *five* digits, instead of *six*?

I confess it seems preposterous to ask such a question as this, when there is more than abundant evidence to show that our own hand and foot, like those of so many mammals, are formed on the

[1] *Darwinism*, p. 389.

THE ONE BIG DIGIT OF THE HORSE 249

same plan. I think the dogma that our hand is made up of five digits only, like that of *Phœnacodus*, is not so unquestionable as might at first appear.

Modifications have been so vast, from the first appearance of a hand and foot, in vertebrates, up to the form with five digits, that it is next to impossible always to discover, in the whole series, which bones are homologous.

In the changes that the carpal and tarsal bones have been subjected to, it is not easy to identify their position regarding the metacarpal and metatarsal bones. For instance, in the extinct *Coryphon* and the *Titanothere*,[1] there is already a displacement of wrist and ankle bones. If in these primitive animals the change of position is already considerable, what shall we say of their position in the Horse and Ox? and in the tarsal bones of Man the change of position is vast.

In the higher mammals, where these limbs have not been much modified in structure, of course the homologies of bones may be easily ascertained. But even here it is not always easy to do so.

Flower and Lydekker[2] give the right hand of a water Tortoise (*Chelydra Serpentina*), Fig. 94. It is an animal far removed from mammals, but nevertheless it has five digits in its hand, and is evidently related to the human hand. It has a separate carpal bone to each digit, all in one row. This order becomes much modified in the carpus of other animals, so much so that Chauveau[3] says: 'The human hand having five digits and five metacarpal bones, it is rational to admit the virtual existence of five pieces to each of the carpal rows. Materially, there are only four bones in each of the rows; but the comparative study of the relations of

[1] *Royal Natural History*, p. 152.   [2] *Mammals*, p. 48.
[3] *Anatom. Comp. des Animaux domestiques*, tr. by Fleming, p. 123.

each of these bones in the human carpus, and in that of animals which are in possession of the archetypal hand, leads to the belief that the *scaphoid* is the result of fusion of the fourth and fifth bones of the upper row, and the *unciform* the fusion of the first and second bones of the inferior row.' This may be so, but let us see what Sir W. H. Flower[1] has to say on the homologies of carpal bones.

'The determination of the homologies of the carpal bones of the Cetacea with those of other mammalia is beset with difficulties, and has consequently led to some differences of opinion among those anatomists who have attempted it. Moreover, every species appears liable to certain individual variations, and sometimes the different sides of the same animal are not precisely alike, either in arrangement or even the number of the carpal ossifications. . . . In many cases, as in the round-headed Dolphin, the bones of the distal row of the carpus are reduced to two, which appear to correspond best with the trapezoid and unciform, the magnum being either absent or amalgamated with the trapezoid. The trapezium appears never to be present (in the Cetacea) as a distinct bone, although the first metacarpal so often assumes the characters and position of a carpal bone that it may easily be mistaken for it.'

Further, some species of Woolly Spider Monkeys (*Eriodes arachnoides*) have a rudiment of a thumb on one hand, and not a trace of one on the other.[2]

Note that '*sometimes the different sides of the same animal are not precisely alike, either in arrangement, or even the number of the carpal ossifications.*'

If the two sides of the same individual are liable to variation in

[1] *Osteology of Mammals*, p. 301, 3rd Edit.
[2] *Royal Natural History*, vol. i. p. 158.

THE ONE BIG DIGIT OF THE HORSE 251

arrangement and *number*, what shall we say of the liability to variation in individuals and races, separated from each other perhaps by millions of generations, and subjected to different nervous influences, and to the action of different surroundings?

It is evident to me that in many instances homologies are rather assumed than proved; and it will be seen how difficult it may be to determine in all cases what particular bone in the hand or foot of one animal is homologous with another bone in the hand and foot of another animal.

Therefore we cannot always be 'cocksure' that in these cases our inferences are beyond question.

Professor Huxley says, 'In matters of the intellect do not pretend that conclusions are certain which are not demonstrated or demonstrable.'

We have now to face four cases of abnormalities, which may interfere with the notion that there is, as Chauveau and others think, such a thing as an archetypal hand or foot with *five digits*. These are as follows:

A. In Chauveau's work[1] are given two instances of abnormal hands in Horses, a sketch of which is shown in Fig. 90. There can hardly be much doubt that these so-called monstrosities are only reversions (at least they *might* be so interpreted) to some very remote type from which the Horse descended, the Ox and its congeners having continued that remote line of type up to the present day.

. We might suppose that while the hand and foot of the Horse have become consolidated throughout into one digit, those of the Ox have become consolidated only as far as the metacarpal and metatarsal bones, leaving the phalangeal portions of the two digits *free*.

I do not see how we can escape from some such conclusion as

[1] *Anatom. Compar. des animaux domestiques*, p. 130.

this, seeing that Fig. 90 (*b*) could stand for the hand of an Ox or Ram. As shown in (*c*), it does not materially differ from it.

A third case of cloven hoof in the Horse was pointed out to me by Professor M'Fadyean of the Royal Veterinary College, and

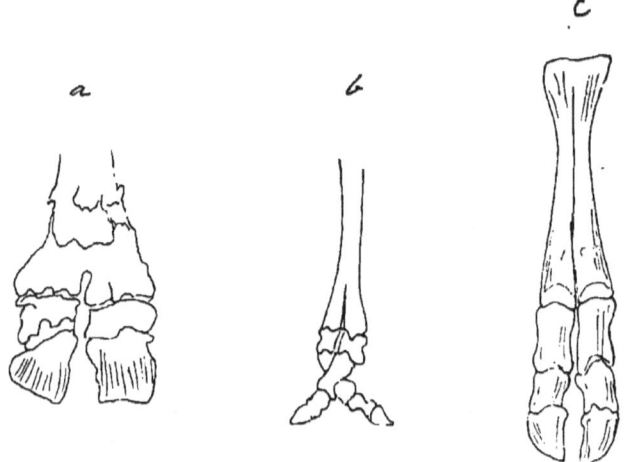

FIG. 90.—(*a*) Abnormal hand of a Horse; (*b*) abnormal hand of a Colt, split into two digits, like that of a Ram, from Chauveau's *Anat. Comp.*, p. 130; (*c*) normal hand of an Ox, from Professor M'Fadyean's *Compar. Anat. of Dom. Animals*, Fig. 87.

illustrated in the *Journal of Comparative Pathology and Therapeutics*, vol. iii. pt. 3, September 1890, p. 263.

B. The fourth abnormality we have to face is this: All osteologists know that long bones have a nutrient foramen for the passage of an artery into the interior of the bone. Well, Professor M'Fadyean, in his work on *The Horse*, on p. 153, Fig. 113, gives the left metatarsal bone of a Horse with *two* nutrient foramina where generally there is only *one*.

THE ONE BIG DIGIT OF THE HORSE 253

Even if there were *always* only one nutrient foramen in this bone of the Horse, it would be no evidence that the Horse's metatarsal bone is not *genetically made up of a fusion of two*, for all the Ox family which I have examined in the Natural History Museum have only *one nutrient foramen in their metatarsal bones, although they are admittedly made up of a fusion of two distinct bones* ! Therefore this case of two foramina given by Professor M'Fadyean is the *more remarkable and important.*

This abnormality and those cloven Horse's hands quoted from Chauveau's work, and by Professor M'Fadyean, render it questionable whether the Horse's digit is merely an enlarged one digit (the iii.) as was supposed.

I think that these teratological facts induce at least a *suspicion*, if not a conviction, that the Horse's hand and foot may after all be double organs in origin, like those of the Ox, fused into one, not only in the metacarpal and metatarsal portions, as in the Ox, but also in their phalangeal portions, that is, throughout, so as to produce *one* solid hoof.

If the Horse's metatarsal bone given by Professor M'Fadyean is not a compound of two bones, then I do not see what business it has with *two* nutrient foramina. I would ask,—Why did two foramina come there at all, when one is the normal state of things in Horses, and when one is quite sufficient for the two fused bones of the ruminants I have mentioned, *if there were no reminiscence of two bones in origin*?

You might ask,—If, as you suppose, the common bone of the Horse may be a fusion of two, why does it never disclose the line of suture, as in the Ox ? Well, we do not know that in the millions of Horses which have been bred no such disclosure has occurred ; all we know is that it has not been noticed and recorded except in cases of Cow-foot. Moreover, there are many skeletons of ruminants in the

Natural History Museum, the cannon bones of which have no trace of suture. Professor Marsh, who has devoted much attention to recent polydactyle Horses, gives a figure of the horned Horse of Texas.[1] That this particular polydactyle Horse was also *horned* may be of some importance in comparing the limbs of Horses and Oxen.

Professor Marsh mentions that Julius Cæsar 'used to ride a remarkable Horse, which had feet that were almost human, the hoofs being cleft like toes.' He gives several cases of additional digits in the Horse, besides the usual one big digit. But none of these seem to be of the nature of the Cow-foot of Chauveau and M'Fadyean.

The size of a digit would naturally depend to a large extent on the fact of the weight of the body being thrown principally on that particular digit. In the Tapir, Rhinoceros, and Elephant, it was thrown on the third; in the Horse, if uneven-toed, also on the third, but if even-toed, like the ruminants, the weight was thrown equally on the third and fourth. In man it is thrown on the first of the foot.[2] In *Chæropus Castanotis*, a marsupial, we have in the hand the third digit largest, while in the foot the fourth is largest and the rest more or less atrophied.

It must not be supposed that a two-digited Ox cannot run as quickly as a one-digited Horse. When in Kandy (Ceylon), I one day hired a wagonette with one large Horse to take me to Peradenya. The Horse trotted all the way. Behind us was a small Zebu Ox harnessed to a native cart. It trotted all the way, and always kept the same distance behind us. Rich natives in India keep large Hansi Oxen, which are very good trotters. I should say there are not many Horses that can gallop as fast as an

[1] *American Journal of Science*, third series, vol. 43, 1892, p. 344.
[2] Some men wear the heels of their boots on the *outer side*, which means that they walk somewhat like a Gorilla, and throw their weight on the *outside* of their feet, and therefore on the *smallest* of their toes.

THE ONE BIG DIGIT OF THE HORSE 255

Antelope; and Dogs with five toes in front and four behind can gallop as fast as Horses.

Then there are so many cases—whole families sometimes—quoted by Darwin, Guinard, and others, of which I have given details in another place, who had *six digits in hands and feet.*

If there be such a thing as an archetypal hand and foot with five digits, where did these numerous persons, and other animals, get their sixth digit from?

The theory of Flower and Lydekker, and others, that the one digit of the Horse is derived from the big middle digit of *Hipparion*, and this from that of *Anchitherium*, and this from the big digit of *Pachynolophus*, and this again from the enlarged middle digit (the iii.) of a *Phœnacodus*, would seem to be plain sailing. But one would like to ask again, Are the hand and foot of *Phœnacodus* composed of five digits or of six, with the middle one a *fusion of two*?

This question would seem so preposterous that it almost takes one's breath away to pen it! Yet teratological facts leave us no alternative but to ask it. Facts are so stubborn that they give one no peace till they are somehow explained.

When we see that the Leopard of to-day still carries on its skin the impressions of plate-armour with which its ancestors millions of years ago may have had their bodies protected, the notion that the Horse's foot is an amalgamation of two ancestral digits which may have occurred millions of years ago does not after all seem so startling. Just think of the immense family of orchids. Some incredible time ago their ancestral pistil and stamen became fused and transformed into the modern orchid-*column*. Yet through the innumerable variations that this curious flower has undergone, from the size of a large pin's head to that of a tea saucer, this characteristic feature has *persisted*. And it is only now and again, by some teratological specimen which happens to turn up, that the secret of its origin is revealed.[1]

The question is whether there is any evidence to show that the Horse and the Ox may be more *closely* related than the Horse is to the Rhinoceros. I think there is. There is, however, no doubt that all three, the Horse, the Ox, and the Rhinoceros, at some still remoter period came from the same stock.[2]

It might be said that nothing is easier than to identify the thumb or big toe, which have only two phalanges, while the other digits have three. Even in the foot of the Seal (*Macrorhinus Leoninus*),[3] where both the thumb and little finger are about of equal length, and both prolonged beyond the other three digits, the

[1] See Dr. Maxwell T. Master's *Vegetable Teratology*, pp. 380-384.
[2] In *Creatures of Other Days*, pl. 21, Mr. Hutchinson has clothed *Phænacodus* with longitudinal stripes somewhat like a *Paca*, or like the *common American Tapir*; and *Hyracotherium* with stripes like a Zebra. For reasons given in another place, it is more probable that these two ancient mammals were *dappled*, somewhat in the manner of a Leopard. As delineated, they are too much like *modern* striped animals. But as they were, as has already been suggested, nearer to their ancestral Glyptodontoid forms, the markings in those early days could hardly have been so modified as those of the Zebra and Tiger of *these* days. They seem to have been given a *recent* dress over an *ancient* skeleton.
Then on pl. 23 Mr. Hutchinson gives a Sabre-toothed Tiger attacking a *Macrauchenia*. This carnivore is shown as striped, like the existing Tiger. Now, the striping of the Tiger, as I have endeavoured to prove, is a much modified rosetting. It does not at all follow that, because the skeleton of the sabre-toothed carnivore is like that of a Tiger, *therefore* its skin-marking was like the Tiger of our days. It is more likely that it was rosetted like a Jaguar, not only because it was *nearer* the basis of its own evolutionary branch, but also because it comes from the *Pampas formations*, from which region the Jaguar also comes. Curiously enough, in the *Saturday Review* of 15th September 1894, p. 306, in a review of Mr. Hutchinson's *Creatures of Other Days*, I find this:—' Here are represented two ancient and ancestral mammals, the *Phænacodus* and the *Hyracotherium* : the former is ornamented with longitudinal stripes, and the latter with transverse bands like a Zebra. Still it is better to have done this than to have taken the greater liberty of making them spotted.'
The reverse is not improbably the case, considering that such a large number of American mammals are spotted, including a large proportion of the fawns of the American Deer.
[3] Given by Professor Flower in *Osteology of Mammals*, p. 347.

THE ONE BIG DIGIT OF THE HORSE 257

thumb with only two phalanges can be readily made out. True, but in this discussion I am taking into consideration not only a broad evolution reaching to the most ancient forms in which we can clearly identify what we call a hand and a foot, but also *teratological*, or monstrous specimens. Now and again these peep out of oblivion to reveal something of the invisible past, and therefore cannot be ignored in framing evolutionary theories.

Although the thumb and big toe are readily identifiable in series of Mammals, from their having only two phalanges, it would appear that the thumb before it got shortened and partially atrophied, and separated from the other digits, also had three phalanges, like the other digits; for in Fleming's translation of Chauveau's work [1] an abnormal hand of a Pig is given (Fig. 89 of these pages), in which the five digits are present, and the thumb is shown with *three* phalanges:[2] moreover, its metacarpal bone is amalgamated with the trapezium of the carpus.

Then in the Ichthyosaurs there does not seem to be any distinction between the thumb and the other digits; and when we consider the large number of phalanges there is in the hand and foot of these extinct animals, we begin to see that homology, beyond a certain stage, cannot always be traced simply from the *number* of bones in a limb. The carpal bones in their descent have undergone so much alteration and displacement that no accurate conclusion can be drawn from this source in all cases. Even in Monkeys, the majority of which have five digits, there is a section with *eight*, and another with *nine* carpal bones.

Why this so-called archetypal form of limb gave rise to such a vast series of five-digited animals we cannot tell. We may, however, assert with some confidence that it was not the five digits

[1] *Comparative Anatomy of Domestic Animals*, p. 122.
[2] Supposing this not to be an oversight.

R

which gave the ancestral forms pre-eminence, and enabled them to stamp this feature on millions of descendants, but some other anatomical and physiological characters, developing *concurrently* with the five-digited hand and foot, which enabled that type to endure and fight its way to our times. Four, three, two, and perhaps even *one* digit were enough to secure survival. It would seem that it was not so much the limbs as the *brain and viscera* which gained the day, in spite of the diminishing number of digits.

It seems impossible to account for atrophy of one or more bones, which may sometimes occur all of a sudden, and give rise to what is called an anomaly. If the abnormality or monstrosity should happen to be useful, or can suit itself to surrounding conditions, it may endure, and be reproduced, and possibly lay the foundation of what might subsequently be called a *normality*. As an abnormality, the animal may have originally been better off in the struggle for existence. Certainly the struggle must be more keen among those who have exactly the *same* structure.

I remember seeing in Jeypore, Rajpootana, an adult who had no arms at all. In what corresponded to the glenoid cavity of the scapula there was a small finger, or at least what looked like one.

Here then was a case not merely of atrophy of four digits with their metacarpal and carpal bones, but suppression of the whole radius, ulna, humerus, etc. Of course, such a monstrosity as this was anything but useful, and not likely to endure.

On the occasion that Professor M'Fadyean showed me the case of cloven foot in a Horse, in the *Journal of Comparative Pathology and Therapeutics*, already mentioned, he remarked that the phenomenon was not very uncommon. The abnormality he showed me was in a Horse four years old, and in only one foot.

The Professor did not seem to look upon this anomaly as a

THE ONE BIG DIGIT OF THE HORSE

case of *reversion*, because the metacarpal and metatarsal bones of the Horse, unlike those of the Ox, were *single* bones. I remarked that the case, in his book on *The Horse*, of a metatarsal bone with *two* nutrient foramina, threw some doubt on the theory that these bones in the Horse were *originally* only *one* bone. He replied— 'Then, if these bones are made up of a fusion of two, you would have to credit the Horse with having had *six digits* originally.' I said that would exactly correspond to the *six digits* in the hand and foot of numerous persons mentioned by Darwin and others.

I do not think that we are justified in beginning to count from either a *Phœnacodus*, or any other extinct animal previous to it, with five digits to its hand and foot, and then call these 'archetypal' hand and foot, for these extinct animals were not the commencement of the vertebrate type, and behind them there may have been other extinct animals which may have had *six* digits. True enough, we have not yet found fossil animals with normally six digits, but we have found some with *more* than six.

There can be no doubt that certain digits can become hypertrophied, and then, through natural selection, the enlargement might be continued, and even increased; but what I have some doubt about is whether the *Horse's* digit was originally an enlarged *one*, or a fusion of *two* digits.

The abnormal examples given by Chauveau and M'Fadyean, which are like the cloven foot of the Ox, and the example of two nutrient foramina in the metatarsal bone of a Horse, certainly tend to throw doubt on the accepted origin of the Horse's digit as a hypertrophied *one* digit. These abnormalities may be as important, in a sense, as the discovery of the archeopteryx.

When we begin to search lower down in the scale of vertebrates than *Phœnacodus* to try and find out where it got its five digits from, what do we find?

260 STUDIES IN THE EVOLUTION OF ANIMALS

In the Natural History Museum there is a fine collection of Ichthyosaurs of the Mesozoic period. Nobody can doubt that their four limbs, however rudimentary, as judged by our own hands and feet, are the homologues of the four limbs of the higher vertebrates. Well, in these Ichthyosaurs the digits of the hands, or what correspond to digits, vary from four to eight. In several instances one of the digits appears as if it were dwindling off. The carpal bones are hardly distinguishable from the metacarpal and phalangeal bones, which are very numerous, showing that in the higher vertebrates the phalanges are reduced in number, either by suppression or by coalescence. What is very notable is the fact that in some the hind-legs are comparatively small, and with a reduced number of digits.

This may perhaps be the reason why some of the higher vertebrates, such as the Tapir, Rhinoceros, and usually the Dog, etc., have a reduced number of digits in *their hind limbs*. There certainly does not seem to be any good reason for this reduction which can be drawn from usefulness and natural selection.

Here then in this Ichthyosaur plane of life we have a basis out of which *several* archetypal hands and feet could have arisen. It only needed that each monstrosity should have been repeated through heredity, and made permanent by survival, as a race fit to cope with its surrounding difficulties.

In the Plesiosaurs, which apparently are closely allied to the Ichthyosaurs, although they have a much longer neck,[1] we find that the hand had already settled down from the many-digited to the five-digited type, as shown in Fig. 93. And there would seem little room for doubt that the Plesiosaur hand, or one like it, was

---

[1] Some of the Tortoises have a much longer neck than any of the Crocodiles; and in the snakes it is not easy to say where the neck commences and where it ends; yet all three sections are closely allied.

THE ONE BIG DIGIT OF THE HORSE 261

the ancestral form of the Dolphin's hand (*Globicephalus melas*) as delineated in Flower's *Osteology of Mammals*, p. 302, in which two of the digits remain almost Plesiosaurian, and the remaining three are vastly modified by atrophy.

In this Round-headed Dolphin, the carpus has become differentiated from the metacarpus and phalanges with a probably greater mobility of the hand. Here is a comparison of the hands of two Dolphins, showing the number of phalanges, including the metacarpals, as given in Flower's *Osteology of Mammals*, p. 303 :—

	Globicephalus melas.	*Delphinus.*
thumb,	four	two
index,	fourteen	ten
middle finger,	nine	seven
ring finger,	three	three
little finger,	one	one

So that in closely allied animals the number of phalanges can be vastly different.

It is not known whether the Ichthyosaurs and the Plesiosaurs were oviparous or viviparous. Hawkins, in the book of the *Great Sea-Dragons*, p. 18, says—'We had hitherto supposed these Sea-Dragons oviparous, and now we are tempted to think them mammal.'

Judging from the skeleton of the hand one would be inclined to infer that, like the Dolphin, the Plesiosaur, in spite of its long neck, was a mammal.[1]

When we begin to compare the hands and feet of vertebrates, it is astonishing what modifications they have undergone. For

[1] In reality the distinction between oviparous and viviparous animals is only nominal, for in the Australian Platypus we have an oviparous mammal.

instance—the Alligator (*Caiman latirostris*) has only four digits in the hind-limb, thus furnished:—

1st digit,	two phalanges,
2nd digit, . .	three phalanges,
3rd and 4th digits, . .	four phalanges each.

Then in the Brazilian Tortoise (*Testudo tabulata*) the hind-limb is as follows:—

1st digit, ...	two phalanges,
2nd, 3rd and 4th digits,	two phalanges each,
5th digit, . . .	may be one, or may be two;

and all the bones of the first row of the tarsus are lumped into *one* bone, so that the modifications in the foot of this Tortoise have been very great.

If we turn to the birds, we find the modifications in their hands and feet still more astonishing.

Among mammals, again, we find a curious modification in the hand of the Little Ant-eater (*Cycloturus didactylus*). The third digit has only two phalanges, a stout and broad metacarpal bone, and the distal row of the carpus is often lumped into *one* bone.

These are a few examples of animals in which the bones of the hand have become so modified as to make it very doubtful whether all their homologies can any longer be traced. The study of the young animals where the ossification may be still incomplete will no doubt help in discovering their homologies. But it requires great faith in the acumen of a professor to believe implicitly all he may write or say regarding homologies; for in this little Ant-eater's hand[1] the bone marked (v.), as the rudiment of a fifth digit, may be one of the carpal bones, and that marked (tm), as the trapezium, may be the metacarpal bone of the first digit!

[1] Fig. 108—Flower's *Osteology of Mammals*.

THE ONE BIG DIGIT OF THE HORSE 263

Atrophy and subsequent suppression, enlargement and fusion, are some of the factors of modification in the skeletons of vertebrates; and we have seen that in the hand of the Plesiosaur and the Round-headed Dolphin there is no distinction between the metacarpal bones and the phalanges.

The more one makes researches to discover at what period the so-called archetypal hand with five digits took form, the more one loses faith in there ever having been such a thing. From the hand with many digits the hand with five digits evolved, and this continued to lose other digits till only one has remained, as is supposed, in the Horse. All that can be said in favour of an archetypal hand is, that a large number of vertebrates have five digits in their fore-limb, and that probably they evolved from a many-digited type *after* it had lost several of its original digits, and was temporarily reduced to *five*. We have seen that some of the specimens of Ichthyosaurs in the Natural History Museum have as many as eight digits in their hand.

I suppose there is no osteologist who believes that this so-called archetypal hand was created *ab initio* with five digits!

It is not only in extinct animals, but also in recent ones, that we find a large number of phalanges—as shown by Professor Flower in the two fingers of the Round-headed Dolphin. Curiously enough, although this mammal has retained some of the general characters of the Plesiosaurian hand, it has got rid of its legs entirely. We have seen that in the Ichthyosaurs the hind-limbs were already *dwindling*. We cannot know which set of extinct animals eventually branched off into the mammal type, and we have no means of knowing whether that immense series of extinct animals, provisionally grouped under the name of *Dinosaurs*, were oviparous or viviparous. Indeed, as I said, this distinction is not of much importance, seeing that viviparous means hatched inside the womb and then brought forth, and oviparous means first brought forth and then hatched outside. We know that there are both oviparous and viviparous fishes. We have some indication that this so-called archetypal hand or foot may have commenced as low down as in the fishes; for Fig. 91 (*a*) would seem to indicate distinctly, in this Fish's *foot*, the beginning of a limb with five digits, which is not very different, externally at least, from the foot of the mammal, (*b*). Then the pectoral fins of shore-fishes and land-fishes, with six or more digits, or what would correspond to digits, as shown in (*c*), (*d*), (*e*), are very suggestive. In some vertebrates we call these limbs pectoral and ventral fins, while in others we call them hands and feet; many zoologists, however, consider them homologous.

Few, I imagine, would deny that the pelvic fin of *Raia clavata*, shown in Fig. 91 (*f*), does not foreshadow the hand and foot of the higher animals. Between the two outer big digits there are many embryonic digits, some of which are not at all dissimilar to the dwarfed digits of the Kangaroo foot (*Macropus Bennetii*), shown in Fig. 122 of Sir William Flower's *Osteology of Mammals*. Whether the minute digits of this Kangaroo have been degraded from a fully developed five-digited limb, or whether they originated from some fish-like limb, and have never been promoted into useful digits, I leave for others to decide.

The duck-billed Platypus is a marsupial of very low type, and a good swimmer. Is there any good reason for supposing that it could *not* have been evolved from some fish-like vertebrate, with fins like those of the land-fish? We know that certain fish can breathe both by gills and by lungs; and we know that the Frog, in its early life, is a fish-like animal, breathing by gills, and in its later life it is amphibious, and breathes by lungs.

It is not easy to get rid of the notion that in this *Lophius* and

THE ONE BIG DIGIT OF THE HORSE 265

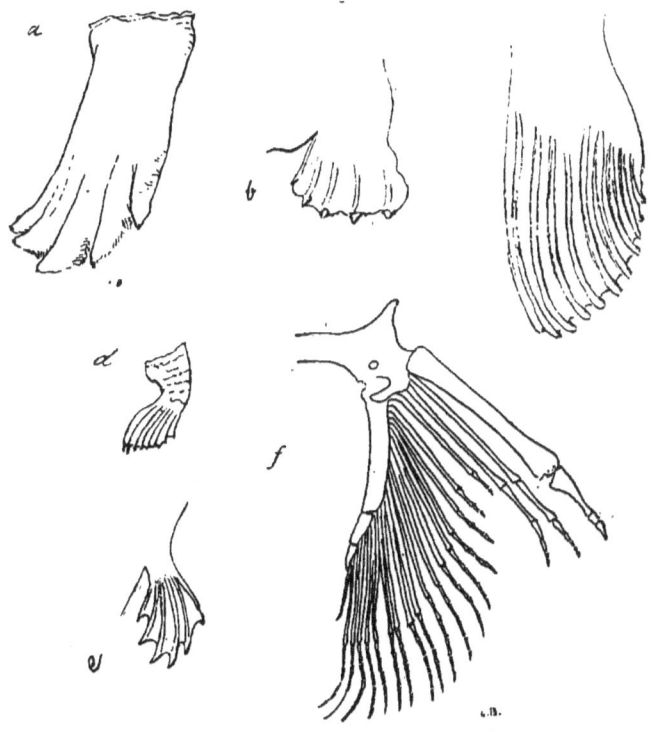

FIG. 91.—(a) Left pelvic fin of *Lophius piscatorius* (Natural History Museum); (b) hindfoot of an Otter (*Enhydris Lutris*), *Proc. Zool. Soc.* 1865, p. 100; (c) foot of a shore-fish (*Lophius Naresii*), *Challenger Exped. Zoology*, vol. i. pl. 25; (d) hand of land-fish (*Periophthalmus*), *Marvels of Animal Life*, by Ch. Fr. Holder, pl. v; (e) limb of a shore-fish (*Zanclorhynchus spinifer*), *Challenger Exped. Zoology*, vol. i. pl. 8; (f) fin of *Raia clavata*, right pelvic (Natural History Museum).

Raia, we have Nature's first tentative rough models of our own hands and feet!

The present classification of vertebrates would be hardly possible if *all* the extinct mammals, and all intermediate forms, which have been either naturally selected out of existence or destroyed by earthly catastrophes, were present before the mind of the zoologist; for the whole series would be more or less a continuous chain without the distinct walls of division or separation, which sometimes form gaps in the present classification.

Professor Agassiz [1] truly said: ' I have already stated that classification seems to me to rest upon too narrow a foundation when it is chiefly based upon structure. Animals are linked together as closely by their mode of development, by their relative standing in their respective classes, by the order in which they have made their appearance upon earth, by their geographical distribution,[2] and generally by their connection with the world in which they live, as by their anatomy. All these relations should therefore be fully expressed in a natural classification; and though structure furnishes the most direct indication of some of these relations, always appreciable under every circumstance, other considerations should not be neglected which may complete our insight into the general plan of creation.'

The upshot of all this discussion about the Horse's big digit is to create a suspicion that it is possibly *not* an enlargement solely of an ancestral *third* digit, but a fusion of the *two* digits of some Ox-like ancestor.

We know that the Ox—an animal seemingly much allied to the

[1] *Contributions to the Natural History of the United States*, quoted by Dr. C. F. Holder in *Life and Work of L. Agassiz*, p. 183.

[2] Geographical distribution often *unlinks* animals. It is only by supposing some means of translation that two distantly located animals of the same or similar species can be considered as having had a common origin.

THE ONE BIG DIGIT OF THE HORSE 267

Horse, as I shall show—and its congeners have the metacarpal and metatarsal bones *fused into one* 'cannon' bone, while the phalangeal portions are separate. We know also that the Horse sometimes reproduces the separation of the phalangeal portions in what is commonly called the cloven-foot or Cow-foot. It is true that in ruminants the cannon-bone has usually a groove which indicates the fusion of two separate bones ancestrally, as shown in Fig. 90 (*c*); but the abnormal Cow-foot of (*b*) shows a similar groove in the lower portion of the homologous bone.

In support of the suspicion that the Horse's *one* big digit may possibly be a *fusion* of the two digits of an Ox-like ancestor, I shall now mention certain other features which are common to certain ruminants and also to Horses.

(*a*) In another place I have mentioned that the Horse has frequently a white blaze, and white hands and feet. These features are almost exactly matched in the African Antelope (*Damalis Pycarga*), the Bonté Bok.

(*b*) The maculations of the Giraffe would appear to be *fusions* of maculations similar to those of the Horse; but the maculations of the Zebu of Fig. 58 scarcely admit of a doubt that the markings of the Zebu and those of the Horse of Fig. 36 are identical.

(*c*) The brindled *Gnu* is striped much like a Zebra, and their manes are very similar. Moreover, I have often seen Horses faintly brindled on the neck almost exactly like the brindled Gnu. Then the Gnu's tail is more like that of a Horse than that of other ruminants. On one occasion I was watching the white-tailed Gnus in the Zoological Gardens. One of them pretended to shy at nothing, just as a young Horse often does.

So that in the Gnu we seem to have an intermediate form between the Horse and other ruminants. It has a short mane, like that of the Zebra; it is striped not very unlike a Zebra or Quagga; it has long hair under its chin, which many cart-horses have; its tail is like that of a Horse; the white-tailed Gnu shies exactly like a Horse; but this Gnu has a tuft of hair over its nose which is unique, and may be somehow related to the horn on the nose of the Rhinoceros, and which is stated to consist of an agglutination of hairs.

The reader might say that the likenesses mentioned may perhaps be conceded, but what about the horns of the ruminants?

(d) Well, in the Natural History Museum it is stated that 'The earliest known forms of Deer, those of the lower Miocene period, had no antlers, as in the young of the existing species. In the existing genera, *Moschus* and *Hydropotes*, there are no antlers at any time. The Reindeer has them in both sexes, and in all others the male only.'

So we see there is a very great variation in these features, and some existing genera among ruminants, and also among extinct forms of Deer, are *without* horns.

Further, let us see what the great naturalists say about horns in the Horse.

(e) Darwin tells us,[1] 'In various countries horn-like projections have been observed on the frontal bones of the Horse; in one case, described by Mr. Percival, they arose about two inches above the orbital processes, and were very like those of a calf from five to six months old, being a half to three-quarters of an inch in length.' And further on, Dr. Wallace says 'That horns have not unfrequently arisen from such apparently uncaused variations is indicated by the remarkable difference of structure and growth in the horns of such allied groups as the Deer and the Antelopes, which at a quite recent epoch must have originated independently.'

[1] 'Are individual acquired characters inherited?' by Alfred Russel Wallace, *Fortnightly Review*, April 1893, p. 495.

THE ONE BIG DIGIT OF THE HORSE 269

With due deference to Dr. Wallace's great authority on these matters, I would ask why does he say 'That horns have not unfrequently arisen from such apparently *uncaused* variations'? They may be *inherited* reversionally from some ancestral form. Moreover, why does he think that the Deer and the Antelopes 'must have originated independently'? Everything we know about these animals tends to the conviction that they did *not* originate independently. The horns of the Antelope have a core covered with hard epidermis; while the horns of the Deer have a core covered with velvety epidermis which is *shed*: the Giraffe retaining a covering of skin on its horns *throughout*. In the Deer the epidermis peels off and leaves the core *nude*. In the Antelope the epidermis hardens into horn, and sheaths the horn permanently, so that the antler of the Deer may be the homologue of the *core* of the Antelope's horn.

The remarkable part seems to be the shedding of the Deer horns. We have, however, an analogous phenomenon in the shedding of plant leaves, in the shedding of oak branches (*Quercus Robur*), in the shedding of the arms of star-fishes.

Then taking into consideration the digestive characters of both groups, the characters of their limbs, and the absence of incisors from the upper jaw of both, it seems difficult to escape from the conclusion that Antelopes and Deer are closely allied, and have originated from a common stock.

A number of common characters in the Horse, normal and abnormal, such as the 'Horned Horse from Texas,'[1] would indicate that it also is a branch of the *same common stock* from which Antelopes and Deer arose.

(*f*) Then the callosities on the Horse's legs are features which the South American Camel-like ruminants, as I have shown else-

[1] *American Journal of Science*, 3rd series, vol. 43, 1892, p. 344.

where, still retain, and for which several other ruminants have substituted tufts of long hair, or hair differently coloured from the surrounding surface. Curiously enough, Grevy's Zebra has lost the callosities from its legs entirely, without leaving any trace of them, while the other Zebras and the Ass have them only on the fore-legs, and the Horse has them on all four. So that the total disappearance of callosities in many ruminants is no evidence whatever that their ancestors never had them. I have not seen or heard of any mention or trace of such a feature in Rhinoceroses, Pigs, etc., to which the Horse is supposed to be closely allied.

The ruminating apparatus of the Ox and its congeners may appear a difficulty in the endeavour to approximate the two groups, those of the Horse and of the ruminants; but we know that there is a section of Monkeys which have a stomach with several diverticuli, and, curiously enough, this form of stomach seems to have some relation to the absence of cheek-pouches, as if these acted somewhat homologously with the *crop* of the bird and the *rumen* of the ruminant. After all, these three features are diverticuli of the digestive tube.

We know that the Horse takes more time to masticate his food than the Ox, and therefore does not require a ruminating organ. It is not difficult to imagine that the ancestors of the Horse at one time lived in a tract free from enemies, and therefore chewed their food leisurely, and eventually got rid of the sacculated stomach, if they previously had it; or reversely, the ancestors of the Ox were in tracts surrounded by enemies, and had to gobble their food and depart as quickly as possible from the feeding-ground; then those who did not die of indigestion gradually got into the habit of regurgitating their food and masticating it leisurely over again. Of course all this was not done in a week.

If one can imagine a Horse evolving out of a fish, there can

THE ONE BIG DIGIT OF THE HORSE

be no possible difficulty in evolving him out of a ruminant. Palæontologists, moreover, will tell you that if you go back far enough in time you will find that not only the Horse and the Ox, but also the *carnivora*, were mixed up in one type. Time is the great factor needed for this evolution, and assuredly there is no lack of it in the universe!

The question, then, finally resolves itself into this:—Is the Horse an 'odd-toed' or an 'even-toed' mammal?

The reader may now judge for himself, and furnish the reply.

Zoologists and palæontologists have done a great deal to link the present with the past, and so have made a surer basis for the doctrine of evolution. But Professor Huxley[1] gives a warning against too hastily concluding that, *because* no evidence of vertebrate animals has been discovered in certain formations, *therefore* none existed during the deposit of those formations, for in p. 88 he says *casts* of bones may be the sole record of animals having existed in a formation. This means that the bones themselves have been *dissolved away*, and have disappeared. If something had occurred to disturb the cast also, no evidence whatever might remain to show that once bones had been buried there.

At p. 90 he says—'From the highest animals, whatever they may be, down to the lowest speck of protoplasmic matter in which life can be manifested, a series of gradations, leading from one end of the series to the other, either exists or has existed. Undoubtedly that is a necessary postulate of the doctrine of evolution.'

But one might ask—Where are the solid evidences of gradations between protoplasmic matter and the higher animals? They are only to be found at present in the human brain as *inferences* from the *known available facts*.

[1] *Lectures on Evolution*—'Science and the Hebrew Tradition.'

MONSTROSITIES AS PROBABLE FACTORS
IN THE CREATION OF SPECIES

s

'It would be well for students of extinct forms of life to enter this domain of science without any preconceived ideas at all! It would save a great deal of confusion and trouble in the end; and moreover would be far more truly scientific. For what right has any one, however great his knowledge or his ability, to dictate to Nature, and to say this or that is impossible—that no reptile, for instance, could possibly have flown; or that such and such teeth were impossible for a reptile? . . . Facts such as those should teach caution, and every student of palæontology will do well to remember the saying of Agassiz: "The possibilities of existence run so deeply into the extravagant, that there is scarcely any conception too extraordinary for Nature to realise."'

Creatures of other Days,
by Rev. H. N. HUTCHINSON, p. 73.

PART XII

MONSTROSITIES AS PROBABLE FACTORS IN THE CREATION OF SPECIES

THE study of monstrous forms, both in the vegetable and animal worlds, is very interesting, because, not infrequently, some important secrets of nature are thereby revealed, which no one would suspect from the contemplation of simple *normal* forms. This study of monstrosities has been dignified by the euphonious term of *Teratology*, and elevated into a branch of the science of biology: but I do not yet see that monstrous forms have taken their place as *links* in the method of creation by evolution.

What is a monstrosity? Darwin says[1]—'By a monstrosity I presume is meant some considerable deviation of structure generally injurious, or not useful to the species.'

And on p. 52, he says—'Under domestication monstrosities sometimes occur which resemble normal structures in widely different animals. If monstrous forms of this kind ever do appear in a state of nature, and are capable of reproduction (which is not always the case), as they occur rarely and singly,[2] their preservation would depend on unusually favourable circumstances. They would also during the first and succeeding generations cross with the

[1] *Origin of Species*, vol. i. p. 51.
[2] We have absolutely no means of knowing that in ancient times they did not occur in numbers.

ordinary form, and thus their abnormal character would almost inevitably be lost.'

In the case of polydactyl monstrosities Darwin himself has shown that this is certainly not the case, the abnormality often seeming to gain strength by *dilution* with the normal form, and Professor Huxley indorsed this view.

It seems we are here dealing with *words* instead of with *facts*; for two metacarpal or metatarsal bones, by fusing into one bone, become a monstrosity, although of a lesser degree, as much as if two eyes fused into one.

We see so many teratological phenomena in man and among domestic animals, originated somehow suddenly that there can be no reason to doubt that in a state of nature similar monstrosities must also have frequently occurred. The majority, no doubt, became extinct from unfitness in the battle of life, but now and again a monstrous individual may have been fitted, not perhaps to carry on the *same* life as those which bred it, but that very monstrosity may have enabled it to acquire a different habit of life, which may have left it *without* competitors, and so enabled to live. Then if it bred with the others out of which it evolved, it would *sometimes* transmit the monstrosity to which it may have owed its life, and so by degrees a new type of animals would come into existence. Having more or less the same habits, they would congregate together and interbreed, and thus *fix* that monstrosity, so called.

Among monstrosities might be mentioned the upper canines of the *Babirusa*, or Pig-Deer, which grow through the upper jaw *upwards*, and curve backwards over the eyes; and what is more, in some varieties they seem to be quite useless, as their points almost touch the skin; so that instead of digging up roots, as other wild pigs do, the Babirusa contents itself with eating fallen fruit.

MONSTROSITIES 277

Mr. Hutchinson evidently did not feel satisfied that the usual *small* variations have been *always* sufficient to account for certain structural phenomena which would seem to leave unexplained gaps behind them in the course of development. For after stating that surroundings must have had a great deal to do with the changes wrought on the structure of animals, he says :[1] 'There is, however, another cause which may be of even greater importance, although it cannot be so easily understood—and that is *internal changes in the animal itself.* All the creatures we see around us are constantly varying, and have done so in the past—now in one direction, now in another. The cause of " variation " is one of the unsolved problems of modern biology or the study of life ; but we do know that a variation occasionally happens to be of such a kind as to make a radical change in the organism, and to fit it for new conditions of life better than its comrades of the same species.' As an example of such a change he quotes the power of 'gestation,' ' whether acquired at once, by some individual, or only slowly brought about after many generations.'

Zoologists have invented the terms 'aberrant forms' to denote forms of animals which do not conform to the theory of 'gradual accumulation of small variations.'

Among fishes there are numerous forms which might have originated in a sudden 'aberration.' For instance, there are two forms of Sword-fish, the ordinary one (*Histophorus*), with its *upper* jaw prolonged into a sword-like point ; and various species of *Hemiramphius*.[2] These have the *lower* jaw prolonged in a sword-like point. Both these forms may have originated suddenly in fishes with equal jaws, either by a monstrous enlargement of one jaw, if both jaws were small, or by a monstrous dwarfing of one

[1] *Creatures of Other Days*, p. 178.
[2] *Fishes of India*, by Francis Day, part iii. pl. cxix.

jaw, if both jaws were originally long. Then such a monstrous weapon having been found useful, it would increase to a certain advantageous maximum, by constant use and direction of nervous energy thereto. There is nothing in such a monstrosity, supposing the abnormal size to have originated in one, to lead us to suppose that these two fishes would have been debarred from procreating with their normal fellows, and perpetuating this abnormality.

Then the Hammer-headed Shark (*Zygæna*) leads to the thought that such an aberration may have originated monstrously by a certain projection of both ocular regions.

Again, the extinct *Dinotherium giganteum* may have had its turned-down lower jaw tusks originating in a sudden abnormal form; and the immense upper jaw tusks of *Elephas Ganesa* (10½ feet long, and massive in proportion) may have become so large by an abnormal growth. Indeed it may have been those enormous tusks which helped to cause the extinction of that particular form owing to their great unwieldiness, and consequent disadvantage in the struggle for life. In other words, they may have increased *beyond* the maximum of usefulness.[1]

[1] In the Temple Flower Show of 1894 there was a *Sedum* ticketed *Sedum virens monstrosum*, consisting of a fasciation of stems covered all over with small fleshy leaves, and resembling a cockscomb (probably *Sedum reflexum cristatum* of Dr. Masters). Its parent form is said to have *trailing stems*. Then there was *Scolopendrium vulgare scalariforme*, said to be a monstrous form of the common Hartstongue fern. There cannot be much doubt that sudden monstrosities of similar and other natures may have originated *new genera*.

One often sees a fusion of twin flowers producing a larger many-petalled and many-stamened flower. There was one such in *Lilium monadelphum Szovitsianum*.

Mr. Heal informed me that he had a monstrous Streptocarpus consisting of a fusion of *four* flowers. He tried to perpetuate it, but its descendants reverted to the one-flower type with five petals of the Gloxinia shape.

No doubt this obliteration of a monstrous type, and reversion to ancestral forms, would often occur in nature, but among millions of such phenomena there might be some which *could* be perpetuated, and we know that six-digitate individuals can be perpetuated *in spite* of dilution with normal blood.

MONSTROSITIES 279

When one reads Professor Flower's book on *The Horse*, one feels a sort of conviction that the method of creation has been modification by very slow steps, now increasing the number and size of parts, then diminishing them, and even suppressing them entirely, according to the dictates of surroundings and the struggle for life and love.

When, however, one looks into Stebbing's book on the *Crustacea*, and turns up *Dorippe Dorsipes*, and others with two pairs of legs on their backs, one becomes staggered, and the former 'slow-step' faith becomes much shaken. One then begins to think that the method of creation may not *always* have been by *slow* steps, and that sometimes a monstrous form turned up, which, if it could live and struggle, and become adapted to surroundings, might become the foundation of a race on what we would call 'monstrous' lines.

Then at p. 40 Mr. Stebbing says: 'The theory that all appendages of a crustacean are either legs or modified legs will strike a casual observer as rather strained in its application to the mandibles. That a Crab should adapt the basal joints of a pair of limbs for masticating its food may seem as unlikely and absurd as that a man should have teeth on his elbows and should draw them up in front of his lips for the purpose of biting and chewing whatever he wished to put into his mouth. To prevent all cavilling, however, on this point of the theory, the King Crab, *Limulus*, is so obliging as to ignore the ordinary mouth organs, and to use the bases of its actual walking legs as mandibles.'

It is no wonder that anomalous legs should occur at the depths of 3050 and 2375 fathoms (p. 19). It would be a wonder if anomalies did *not* occur at that depth. It is a marvel that the molecules of the ova can arrange themselves at all under that pressure.

Dr. Alfred Russel Wallace says:[1] 'Herbert Spencer tacitly

[1] Are individual acquired characters inherited?' *Fortnightly Review*, May 1893, p. 657.

assumes that natural selection works by the preservation of large individual variations, 'sports' as they are often termed; whereas both Darwin himself, and all his followers, entirely reject these as causes of modification of species (except perhaps in rare cases where they may initiate new organs) and rely wholly on those individual variations which occur by thousands and tens of thousands in every generation.'

It may be heresy to say so, but it does not *follow* that Darwin and his followers are *right*.

Professor W. Kitchen Parker was a follower of Darwin, yet in his *Mammalian Descent*, p. 93, he says: 'Nature does now and then make amazing leaps, certain types taking on sudden metamorphorses, and in the fraction of a lifetime the low is transformed into the high.'

And Mr. Camille Dareste, another follower of Dawrin, writes:[1] ' Aujourd'hui le plus grand problème de l'histoire naturelle est celui de l'origine des formes innombrables sous lesquelles la vie s'est manifestée à la surface de la terre. Si ce problème est soluble, il ne peut l'être que par la connaissance de la tératologie et de la tératogénie ; c'est-à-dire, par l'étude de toutes les formes nouvelles qui peuvent dériver d'une forme spécifique primitive, et des causes qui déterminent leur apparition.'

Dr. Wallace and others then hold to the notion that species have come about *solely* by *gradual* accumulation of useful characters, and only very exceptionally by what may be called '*jumps*'; yet nothing seems clearer than that 'jumps' have occurred in the past,—and may occur now—which, if advantageous in the struggle for life, may have laid the foundation of new types. Nothing is clearer than the fact that a monstrosity may be inherited, and *survive even when it crosses with normal forms;* and the notion

[1] In a Preface to M. Guinard's *Précis de Tératologie*, p. ix.

that it must be extinguished by dilution with normal blood does not seem tenable.

'Natura non facit saltum' may after all be one of those dogmas of Science which may be upset when we begin to perceive that there is no good reason why nature should *not* make a 'saltum.'

It is a matter of history that the doctrine of Natural Selection was suggested in Darwin's mind by the studies he made in the artificial selection of domestic Pigeons and other animals; but Naturalists seem slow to accept the suggestion which monstrous forms in domestic animals and man would naturally lead to, viz., the probability that similar monstrosities in past ages may have been *factors* in the evolution of animals and plants.

And the reason may be that there lingers in men's minds a theological bias in connection with the *word* 'monstrosity.' How could a perfect God allow such a hideous anomaly to live, and engender others like it? But if He did not allow it, how did it make its appearance?

In the Royal College of Surgeons there is the skeleton of a wild Cat with six legs. It was entrapped when half grown. It is evident that it could have lived up to that age, and have struggled successfully for its existence, and, if not captured, it *might* have laid the foundation of a race with six legs! The extra pair of legs are attached to the pelvis, and are evidently the only remnant of a *twin* Cat.

Professor L. Agassiz seems to have had a notion that the 'Divine thought,' now and again, burst out into a new line of evolution, as if it were tired of amusing itself on the old lines, and wanted to try some new experiment.

Unfortunately for this mode of theology, the new type did not, in most cases, cut itself off from the old lines entirely, but seemed to be made up of the old pattern, with something new added on, or taken off, or something fused with another, so that traces of an evolution of one type from another which may have preceded it are in most cases evident.

Then if the Divine thought of Agassiz was not the director of the evolution of these new types, is there any other mode of explaining the sudden appearance of what may truly be called *monstrosities*?

The Science of Teratology has recorded a great number of monstrosities, both in the vegetable and animal kingdoms. What brings them about?

There seem to be two divisions of monstrosities:

(*a*) Fusion, more or less complete, of *two separate individuals* (twins), into one compound body, with or without suppression of some of the parts; and

(*b*) Fusion of different parts of the *same individual;* separation of originally united parts; displacement or suppression of various parts; or re-appearance of some ancestral character, by what is called *reversion* (atavism); or enlargement or diminution of a part.

We know that in the fully formed individual the nerve-centres, with their communicating nerves, control everything. Every function of the body is controlled by them. It is reasonable therefore to infer that, during embryonic development, as soon as the nervous system has had a commencement, it must begin to control the evolution of every part of the body; and the slightest change, from whatever cause, from the normal that may occur in the nerve-centre, or prime-mover, is likely to be reflected as a *modification from the normal* at the distal end of the nerve or nerves which that nerve-centre controls. A very slight change in the embryo may result in much more extensive changes in the fully formed animal.

It would appear a not uncommon thing for the controlling

MONSTROSITIES 283

nerve-centres to produce *un*symmetry of the two sides. And one can imagine that a slight difference in the circulation of the two halves of the nervous system might cause this want of exact symmetry.

The examination of the two sides of spotted and striped animals, such as those of Figs. 4 and 22, will convince any one that a mathematical symmetry of both sides of an animal, if it occurred at all, must be rather the exception than the rule.

Professor Flower has stated that in the Cetacea, even the two hands may have a *different number of bones*.[1]

In the *Royal Natural History*, vol. ii. p. 143, there is the description of a very interesting phenomenon, because evidently rarely met with. The writer says : 'One of the two known skulls of Ross's Seal is peculiar in that, while on one side the first upper cheek-tooth, and both the corresponding lower teeth, are imperfectly divided by a vertical groove, on the opposite side of the upper jaw the place of this tooth is taken by two complete single teeth. Hence it is obvious that we have here a case where an originally single tooth divides into two distinct but simpler teeth. . . . This serves to show how the numerous simple teeth characteristic of the toothed Whales may have been derived by the splitting up of teeth originally composed of three distinct cusps, like those of the Leopard Seal, each cusp of such tooth forming, as we shall see, a distinct tooth in the Whales.'

The writer mentions other peculiarities in the teeth of the Grey Seal on pp. 134, 135.

All this is very instructive, not only because it shows us that on one side of the same jaw there may be simple teeth, while on

[1] 'Sometimes the different sides of the same animal are not precisely alike, either in the arrangement or even the number of the carpal ossifications.'—*Osteology of Mammals*, p. 301.

the other there may be compound teeth, but also because there may be a 'reverse' to the phenomenon described, viz., that a large complicated tooth, such as we see in certain mammals, may be a *fusion* of a number of *simple* teeth! And this is exactly what may have happened, either suddenly, or by slow steps, when a long jaw with many teeth passed into a short jaw with few teeth. Some of the teeth may have been suppressed, but a number may have become fused into *one large compound tooth*.

We do not know what influences and controls the disposition of the parts *before* the nerve-centres of the embryo are formed. It does not follow, because there are yet no nerve-centres and nerves, that therefore there is not something that controls the cells and acts as a means of communication.

Anyhow, when in the animal embryo the nervous system has once taken form, we cannot have much doubt that it controls and regulates the progress of the embryo. Its controlling influence in the multiplication and disposition or arrangement of the body cells, which are increasing by fission, may be likened to the controlling influence of the battery or dynamo in electro-plating. The two processes would appear to be parallel. We do not know why some metals are deposited in a compact form, suited to electro-plating, and why some are deposited in *dendritic* or tree-like forms. All we know at present is that they *do so*. And all we can say of monstrosities is that they *do occur* during the evolution of the embryo, although from recent experiments in artificial teratogeny we may hope at no distant period to learn something more definite about the intimate causes of monstrosities.

Some might say—'This may be an ingenious stretch of thought, which would somewhat explain aberrations in *animals* ; but what about analogous aberrations in plants? We have no nervous system there.' No, but 'nervous system' may be only *words*

MONSTROSITIES 285

intended to give cover to our ignorance. We call nervous system an apparatus consisting of nerve-cells and nerve-filaments. Plants may not have this, but they may have some tissue—unlike nerve-tissue—which *functions* as communicator of *change*-impulses to distant parts. At all events we *must* infer that *Drosera, Dionæa*, and others, possess some means of sending messages from one part of their body to another. Lauffen in Switzerland is a long way from Frankfort, yet recently an influence of a dynamo at Lauffen was transmitted to Frankfort, a distance of one hundred and eight miles, with the greatest ease, although in this case also there was no nervous apparatus to do it!

It would seem that disturbances may occur either in the cells, *before* the nervous system is traced out, or *after*, in the grey matter of nerve-centres. Professor V. Horsley[1] shows that these centres with their characteristic nerve-cells and nerve-fibres are developed very early in the embryo.

New types may have had their genesis during embryonic life, while great disturbances in geological periods were going on, and which may have been due to external physical causes acting on the contents of the ovum.

Sir J. William Dawson[2] says: 'In the Appalachian region of America we have the carboniferous beds thrown into abrupt folds, their shales converted into hard slates, their sandstones into quartzite, and their coals into anthracite, and all this before the deposition of the Triassic Red Sandstones which constitute the earliest deposit of the great succeeding Mesozoic period.'

This seems evidence enough of the commotion caused by volcanoes and shrinkage, and the changes induced by heat and chemical substances.

At p. 176 he further says: 'At the close of the Permian

[1] *Brain and Spinal Cord*, p. 190.   [2] *Geological History of Plants*, p. 175.

and the beginning of the Trias, in the midst of this transition of physical disturbance, appear the great reptilian forms characteristic of the age of reptiles, and the earliest precursors of the mammals, and at the same time the old carboniferous forms of plants finally pass away to be replaced by a flora scarcely more advanced, though different, and consisting of pines, cycads, and ferns, with gigantic Equiseti.'

A little exercise of the imagination will enable us to comprehend that in those volcanic periods, both the sea and the atmosphere may have had gases and other ingredients in them which might either kill, or *strangely modify*, the development of the ova and seeds on which they acted. Heat, electrical changes, and continual tremors of the earth and sea, may have also been factors in the modification of both animal and vegetable embryos. We know what strange modifications of buds and other parts of plants are produced by even the tickling of the minute larva of a Gall-Fly.

In our present comparatively quiet times, with a settled and steady composition of atmosphere and sea, we may have difficulty in conceiving and measuring the amount of modification that may have occurred when the conditions must have been wholly different.

We should not forget, however, that, although not generally admitted, the action of changes in the environment in modifying forms, already established elsewhere, is of importance.[1]

Some species no doubt are very plastic and can easily change their form under changed conditions; others do not, while there are some that are killed outright.

At a lecture of the Royal Horticultural Society, Mr. Burbidge mentioned that *Nepenthes Raja*, from Borneo, cannot be grown in this country under artificial conditions, even with the skill and scientific manipulations of Messrs. Veitch and Sons.

[1] See *Nature*, vol. 43, p. 581, for change of species in *Arabis anachoretica*.

When I visit the Fish Gallery of the Natural History Museum I feel a sort of conviction that many of the strange forms around me must have *commenced as monstrosities*.

The saw of the Saw-Fish (*Pristis*) seems on the face of it to have been a monstrosity; the Spoonbill Sturgeon (*Polypodon spatula*); the curious nose-process of *Callorhynchus antarcticus*, the tailless *Orthagoriscus oblongus*, the enormous dorsal fin of one of the Sword-fishes, *Histophorus gladius*, and so forth—all convince me that these strange forms very probably commenced by a *sudden monstrosity*. It is amongst fishes that we meet with such startling abnormalities in large numbers. Scientists try to explain the stump tail of a Puppy by calling it an 'arrest of development.' That would mean that, 'it is what it is'! Any boy in the street can tell us that such a Puppy 'has not a long tail, like other Dogs!' But what is the cause of the arrest?

Mr. Stebbing in his *Crustacea* mentions that the male of *Gelasimus arenatus* (De Haan) has one cheliped (claw), *right* or *left*, monstrously enlarged. The very fact that it is sometimes the right, and sometimes the left, which is so enlarged, would tend to show that it is not by slow degrees that it attains that size, but by some *sudden derangement* in the right or left nerve-centre that controls the growth of the cheliped, and the Crab has to make the best use of it.

Monstrosities being congenital might easily be inheritable more or less, provided the reproductive organs were efficient. So this monstrosity in a certain Crab may become the fixed feature of a whole race, which originally may have commenced by a sudden evolution in *one* generation, and not by slow and gradual accumulations of some slight variation. The study of fishes and crustacea would seem to afford quite a revelation, as to the probable origin of many species.

288 STUDIES IN THE EVOLUTION OF ANIMALS

One cannot look at the drawings of fishes, even of a restricted region, such as that of India, by Francis Day, without feeling that monstrous forms must have been generated among fishes in plenty. Agassiz, in the account of his *Journey in Brazil*, gives us a striking tale of the *innumerable* species of fishes he and his assistants discovered in the Amazon and its tributaries.

If experiments in teratogeny be worth anything, they will certainly modify our notions of the origin of species, *solely* by *slow* accumulation of some *minute* variation. Just think of it; the ovum segments into two and then into four cells, and so on. If any injury should occur to any one of these cells, from *whatever cause*, the result may be a monstrosity. The wonder is that we are not all monsters! We say heredity keeps us straight; we might also say it keeps us crooked, to allow for variations; but when these statements are translated into common language they seem to spell—'we know nothing about them'!

A little more atomic disturbance here, a little less there, during the embryonic stage, may produce a *new compound*, which then may be called a species, a genus, or even an order, as the case may be. Naturalists may think they have discovered *all* the species in a district, but new ones may be naturally manufactured now and then, just as the chemist produces new compounds by manipulating and acting upon old ones.

There cannot be much doubt that the action of the nerve-centre, in co-operation with the vaso-motor nerves which affect the nourishment of parts, may cause the degeneration, atrophy, and eventual suppression of an organ.

Let us suppose, for instance, a Dinosaur of the ordinary kind, with fore-legs sufficiently large to enable it to move on all-fours, say, like a Kangaroo. Let us suppose again that by a sudden monstrosity the fore-limbs were dwarfed to the size of those of *Cerato-*

saurus.[1] It is evident that if the animal lived, it would be *forced* to walk *mainly* on its hind-legs, and balance itself by its huge tail. It is also evident that the continual use of the hind-legs alone would strengthen them, and at the same time weaken the fore-legs ; and this might go on till the fore-limbs became suppressed altogether, as in *Hesperornis regalis* ; and as the hind-limbs have been suppressed in the Round-headed Dolphin and in other Cetacea. If the fore-limbs can be *suppressed* altogether at once, as in some cases of human monstrosities, they certainly can be *dwarfed* all of a sudden.

Abnormalities are revelations that cannot be ignored. They indicate to us one possible method, by which important *large and sudden* variations may have been effected.

Then, when a large and sudden variation occurs in the brain itself, we call the possessor of it either an imbecile, or a lunatic, or a genius. If the latter, he may be a prophet, a philosopher, a poet, a scientist, a mechanician, an inventor, a saint, etc.

As long ago as 1829, Isidore Geoffroy St. Hilaire [2] mentioned the artificial production of monstrous birds by acting in various ways on the eggs, and he said that artificial incubation gives rise to more monstrosities than natural incubation. And at p. 71 he stated that the history of human monstrosities has frequently been traced to blows on the abdomen during gestation.

Some better and more accurate idea may be formed of the possible origin of variations, anomalies, and monstrosities—all indicating different *degrees* of disturbance during embryonic life— from the recent experiments of Roux, Dareste, Windle, Hertwig, Driesch, Chabry, and others.[3]

[1] Shown on p. 125 of *Creatures of Other Days*.
[2] *Propositions sur la monstruosité*, p. 70 (*Thèses de la Faculté de Médecine*, No. 185).
[3] 'Experimental Embryology,'—by Mr. J. A. Thomson, *Natural Science*, April 1893, p. 294.

All students of embryology know that the ovum, in developing, divides into two cells, then into four, then into eight, then into sixteen, and subsequently into a number of cells called a 'morula,' and further on a 'gastrula,' after which the embryo begins to take shape.

By these experiments it has been shown that 'the germ is plastic in the grip of its environment, and various malformations have been induced which are of interest to the student of morphology'; that 'there is now good evidence to prove that these disturbing agents act, at least in the majority of cases, on that part of the developing organisation which is concerned with the formation of the vascular system of the embryo';[1] that 'of the first two cells into which the egg of a Frog develops, one has in it the material for forming the right half of the body, the other has in it the material for forming the left half of the body.' It was proved that 'one of the first two segmentation cells may form half an embryo, that it can develop apart from its neighbour, and that either a right or a left half-embryo might be produced, as well as an anterior or posterior half'; that, quite lately, Roux has been able to rear an entire Frog embryo from half an egg'; in experiments on ova of an *Ascidian*, Chabry found that 'by destruction of one of the cells into which the ovum first divides, the remaining cell develops into a half-larva, and if two anterior cells of the four-celled stage be destroyed, a posterior half-individual results'; that 'on to the sixteen-cell stage at least, each cell has a determined destiny, and represents a definite part of the embryo, and that if one of these sixteen cells be destroyed, the defect (or monstrosity) in the larva is a definite one'; that 'while Roux got half-embryos from half an egg, Driesch got half-sized, but otherwise complete

---

[1] Surely the vascular system is under the control of the *nervous* system, or of some electrical influence which corresponds to it, before the regular nervous system is developed.

embryos from half an egg'; that Mr. Edward B. Wilson of Columbia College, New York, 'has produced out of one egg of Amphioxus, twins of half the normal size, and by varying the experiments, a sort of Siamese twins from one egg were produced.'[1]

A number of other experiments on ova are mentioned by Mr. Thomson, and the conclusions he has come to from this review are that:

(1) There is a great deal of life in an egg; three-quarters, a half, or a quarter of an egg will, under favourable conditions, form a complete larva.

(2) There is no little plasticity in the germ; the segmentation may be profoundly altered, the shape of the young embryo may be greatly changed, and a new type of larva may be produced, yet the inherited characteristics are strong, for the experiments show a marked tendency in the germ to reach a normal result by an abnormal path.[2]

With regard to the non-inheritance of acquired characters, he says: 'How many ova are there which float in the sea, and in other media; these are now, as similar ova have been in the past, exposed to the influences of very complex physical and chemical conditions. That their living stuff may be greatly affected the results of experimental embryology show. It is likely that the same is true in Nature's great laboratory, and the results, being germinal, may be transmissible. We need not be in haste to exclude the direct influence of the environment from among the primary factors of evolution.'

[1] If all this can be established, it may go a long way towards demonstrating that our two halves were originally a fusion of *two* embryos with *suppression* of a right and left half of each.

[2] Perhaps the embryo may be likened to a crystal; you may break a crystal in pieces, and each piece, under proper conditions of food and stimulus, will grow again into a crystal like its parent.

We may be on the eve of momentous revelations concerning not only the appearance of monstrous forms among human beings and domestic animals, but also concerning the sudden appearance of *new types* of animals in geological periods on *monstrous lines*! These artificial ways of evolving monstrosities may yet be able to teach us how the long tail of the Archeopteryx may have been reduced *in one generation* to the short tail of the modern bird, or *vice versa*; how the long tail of one kind of Monkey may have been reduced in one generation to the stump of other kinds, and to the diminutive tail of Man; how the many digits of the Ichthyosaur may have been suddenly reduced to the five digits of the Plesiosaur, or of some other ancestor of the numerous tribe of five-digited animals.

'We must not forget,' says Agassiz,[1] 'that we are the lofty children of a race whose lowest forms lie prostrate within the water, having no higher aspiration than the desire for food; and we cannot understand the possible degradation and wretchedness of Man without knowing that his physical nature is rooted in all the material characteristics that belong to his type, and link him even with the Fish.'

Changes in the cells of the ovum during segmentation, changes after the embryo became differentiated through disturbance of the controlling nervous and vascular tissues, might at any time have given rise to what is called a monstrous form, which if unfit would die out, but if fit might endure and transmit its monstrous form.

It would all depend upon whether the monstrous form could enter into sexual union with the normal forms. In the case of most fishes, however, even this would not be needed, and the facility with which the sperm cell can reach the germ cell in these low

[1] *Life and Work*, by Dr. Charles F. Holder, p. 189.

MONSTROSITIES

vertebrates, may be the reason why so many strange monster-like forms are found among these animals.[1]

It might be thought incredible that a monstrous form could pair with a normal form. It would all depend, as I said, upon whether the sperm cell of the one could be made to reach the germ cell of the other. By artificial means Messrs. Veitch and Son have succeeded in mating a *Lælia* with a *Sophronitis*—two forms of orchids which no botanist would have conjectured are mere variations of the *same species*!

This is not all, for Professor Agassiz[2] says, 'I remember to have found in the neighbourhood of Mobile (U.S.), no less than six new species in the course of an afternoon ramble. These fishes are almost all viviparous, or at least lay their eggs in a very advanced state of development of the young.[3] The sexes differ so greatly in appearance that they have sometimes been described as distinct species, nay, even as distinct genera (*Molinesia* and *Pœcilia*). We must be on our guard against a similar mistake.'

If husband and wife of these fishes can be so distinct as to be described as distinct species and genera, and *still pair*, what wonder would it be if monstrous forms, produced by disturbing influences which may have occurred during the first stages of embryonic development, should be able to pair with their normal progenitors, although, may be, quite distinct in anatomical characters?

At one time there was a dogma among biologists that animals of different species could not interbreed, as, if they happened by some chance to interbreed, their progeny was barren. It was thought perhaps that some providential law ruled matters thus, in order to prevent the mixing of species. But now it seems to be

---

[1] The variations of type in the Fishes of the Amazon, noted by Professor L. Agassiz, are simply *innumerable*.   [2] *Journey in Brazil*, pp. 32, 33.
[3] He does not mention how their ova are fertilised.

admitted that Goats and Sheep, Rabbits and Hares, common Fowls and Pheasants, Pheasants of different species, Tigers and Lions,[1] Zebras, Asses and Horses *do* interbreed, and several of them produce fertile hybrids. The question is—What is a species?

What possible influence in ancient times could have caused a disturbance in the development of ova so as to evolve abnormal forms, say in fishes, as evidenced by the vast number of new types linked by some characters, and yet unlinked by others, with their forerunners. Several causes suggest themselves, viz. :—

(*a*) Submarine volcanoes heating the water and diffusing new chemical compounds dissolved in it.

(*b*) Shakings and concussions of the water and of the sea bottom.

(*c*) Electrical discharges connected with submarine volcanic explosions, and so forth.

I shall now discuss some human monstrosities to see how far such a notion as I have foreshadowed in the foregoing lines has any basis in fact.

Cases of supernumerary digits in Man are by no means rare. Darwin[2] states that 'Dr. Burt Wilder[3] has tabulated the cases of 152 individuals with supernumerary digits, of which 86 were males, and 39, or less than half, females ; the remaining 27 being of unknown sex'

Then[4] he says that 'supernumerary fingers and toes are eminently liable, as various authors have insisted, to be inherited. Polydactylism graduates by multifarious steps, from a mere cutaneous appendage, not including any bone, to a double hand. But (p. 458) an additional digit, supported on a metacarpal bone, and furnished with all the proper muscles, nerves, and vessels, is sometimes so

[1] See *Nature* of 27th April 1893, p. 607, on Lion-Tiger hybrids.
[2] *Descent of Man*, p. 223.
[3] *Massachusetts Medical Society*, vol. ii., No. 3, 1868, p. 9.
[4] *Animals and Plants under Domestication*, vol. i. p. 457.

perfect (note this) that it escapes detection unless the fingers are actually counted; occasionally there are several supernumerary digits, but usually only one, making the total number six. This one may be attached to the inner or outer margin of the hand, representing either a thumb or little finger, the latter being the more frequent. Supernumerary digits are more common on the hands than on the feet, but generally both hands and both feet are similarly affected.'

Mr. Darwin adds that 'the presence of a greater number of digits than five is a great anomaly, for this number is not normally exceeded by any existing mammal, bird, or reptile. Nevertheless, supernumerary digits are strongly inherited; they have been transmitted through five generations, and, in some cases, after disappearing for one, two, or even three generations, have reappeared through reversion. These facts are rendered, as Professor Huxley has observed, more remarkable from its being known in most cases that the affected person has not married one similarly affected.'

In such cases, the child of the fifth generation would have only one-thirty-second part of the blood of his first sex-digitated ancestor. Other cases are rendered remarkable by the affection gathering force, as Dr. Struthers has shown, in each generation, though in each the affected person married one not affected;[1] moreover, such additional digits are often amputated soon after birth, and can seldom have been strengthened by use. Dr. Struthers gives the following instance: 'In the first generation an additional digit appeared on one hand; in the second on both hands; in the third, three brothers had both hands, and one of the brothers a foot, affected; in the fourth generation all four limbs were affected.' Dr. Struthers however asserts that cases of non-inheritance are much more frequent than cases of inheritance.

[1] Note this. The affection *gathers force even when diluted with normal blood.*

Six digits have also been observed in negroes as well as in other races. Also in hind-foot of a Newt, and in hind-feet of several generations of Cats.[1] In several breeds of Fowl the hinder toe is double, and is transmitted truly.

Dr. J. W. Ogle gives a case of inheritance of deficient phalanges during four generations.[2]

Mr. Darwin gives references for all he states.

What we have particularly to remember is that monstrosities, although diluted with normal blood, (a) may *increase in number* in subsequent generations, and (b) may spread from the hands to the feet also.

From this we may infer that the disturbing influence in the hand nerve-centre, which causes the monstrosity in that limb, is liable to diffuse itself along the spinal cord and influence the nerve-centre of the foot also.

A few detailed cases will impress on the reader's mind not only this sex-digitate feature in Man, but also the *persistence* with which it may be inherited, although diluted by normal blood.

Godehen in 1751 recorded the oft-quoted history of one Grezio Calleja, a polydactyl.[3] He had six fingers and six toes in each limb. He had four children, three boys and one girl. The eldest boy, Salvator, had six digits in each limb, but the supernumerary digit in the hands was not so well formed as those of the father. On the contrary those of the feet were better formed. The other three children had the normal number, five, in both hands and feet; but all excepting the youngest had deformities of fingers more or less marked. So only one of the four children of Grezio had normal limbs. The children of this latter were normal. On the

[1] *Brit. and For. Med. Chirurg. Rev.*, April 1872.
[2] J. T. Cunningham, translator of Eimer's *Organic Evolution*, says he has a Cat with six toes on every foot.
[3] *Hist. des Anomalies*, by Isidore Geoffroy St. Hilaire, vol. i. p. 699.

MONSTROSITIES

contrary, Salvator had two boys and one girl sex-digitate, and one boy nórmal. Grezio's second son George (who was normal) had three sex-digitate daughters, and one normal boy; and Grezio's daughter (who was also normal) had two boys and one girl nórmal, and óne boy sex-digitate.

Dr. Prosper Lucas[1] gives the case recorded by Maupertuis. 'Jacob Ruhe, surgeon of Berlin, was born with six digits in his hands and feet, inherited from his mother Elisabeth Ruhen, who inherited this abnormality from Elisabeth Horlsmann. Elisabeth Ruhen transmitted it to four children out of eight. The father was Jean Christian Ruhe, who was normal. Jacob Ruhe, one of the sex-digitated boys, married Sophie-Louise de Thingen, a normal woman, and had six children, two of the boys being sex-digitate.'

Renow recorded several sex-digitate families, who from time immemorial were spread over several parishes of Bas-Anjou. These abnormalities were perpetuated in spite of the alliance of the abnormal individuals with families of normal conformation. The abnormalities were transmitted indifferently to either male or female children. It was also established that a normal child of an abnormal parent can again give birth to an abnormal child.

Dr. Prosper Lucas (p. 327) quotes some much more interesting phenomena. He says that Vanderbuch collected records of a Spanish family where this abnormality was often complicated with the following: 'In the majority of the members of this numerous family, the third and fourth digits[2] of the hand were united by the integument throughout their length. The phalanges of such double fingers were almost all composed of double bones, situated side by side. The nails, although of one piece, had a

---
[1] *Hérédité naturelle*, vol. i. p. 326.
[2] Exactly those that form the fused metacarpal bone of the Ox and its congeners. In the marsupials this is a common feature.

vertical furrow; what indicated that they were in origin two fingers was that the tendons ("epicondylo-sus-phalangetticus, et cubito-phalangetticus communis") were also double. In some individuals of this family the thumb was biphalangian, and among these the extremity of the digit was sometimes bifurcated. In other individuals the two portions which formed the digit were united throughout as in other digits.' What is still more curious is that in some individuals the third and fourth toes were similarly bound up in one integument.

Vanderbuch counted forty individuals of this Spanish family in whom some abnormalities of digits existed. Almost all were healthy.

A race of Cats is mentioned [1] with six digits on each foot, the peculiarity having been inherited to the tenth generation.

Then M. Guinard [2] quotes the case of a Bitch with six digits, which transmitted this feature to almost all its young ones. He also quotes Lenglen, who gives a case of a sex-digitate Man, whose descendants, up to the sixth generation, all presented this feature. He winds up by saying—'Everybody knows the history of that Arab tribe of the Foldi, whose children are all born with twenty-four digits. The members of this tribe are very numerous, and do not ally themselves with other tribes. They consider this feature absolutely constant, so that when by chance a woman gives birth to a child with five digits, she is considered to have committed adultery, and her child is not acknowledged by her husband.'

And finally (p. 131) M. Guinard writes—'Mais en face de ces faits, nous comprenons, sans l'admettre cependant, qu'on ait pu se demander si le chiffre 5, adopté comme représentant le nombre normal des doigts, n'est pas un peu arbitraire, et s'il n'a pas existé des individus ayant normalement six doigts.'

Well may M. Guinard ask whether the *archètypal* five-digited

[1] *Royal Natural History*, p. 427.   [2] *Précis de Tératologie*,' p. 130.

MONSTROSITIES

limb is *not a little arbitrary*! I read M. Guinard's *Précis* after I had been long contemplating the possibly fictitious *archetypal* hand and foot which characterise so many animals.

If by archetypal we only mean a type of animals with five digits which is very ancient, then the term is admissible enough, for

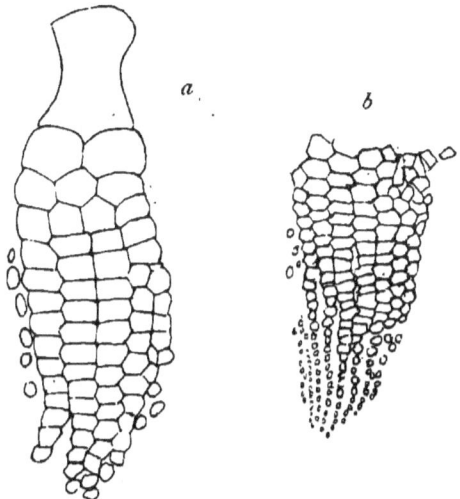

FIG. 92.—(*a*) Right hand of an Ichthyosaur, pl. 23, *Book of the Great Sea-Dragons*, by Thomas Hawkins; (*b*) Right hand of *I. Chiropolyostinus*, digital portion—pl. 7 of same work.

as low down as the Plesiosaurs, and may be much lower, we already find the five-digited type established.

Where could this abnormality of sex-digited animals have come from? Why the five-digited type was so early established we do not know, but we do know that the Ichthyosaurs, which cannot be denied relationship to other vertebrates, had as many as *eight* digits (Fig. 92 (*b*)); while *I. Chiroligostinus* (pl. 3 of Hawkins'

300   STUDIES IN THE EVOLUTION OF ANIMALS

Great Sea-Dragons), had only three complete digits, and a dwindling fourth in both hand and foot.

All we can say is that mammals[1] may have started from a branch of the great vertebrate series in which the abnormality of five digits (we now call this normal and archetypal) had already become a fixed character, and not impossibly the abnormal hands of the Plesiosaurs, with a reduction to only five digits (Fig. 93), and also of their close relatives, the existing Dolphins,[2] may have been the monstrous foundation of the whole race of animals with five digits. We cannot for one moment suppose that in the Ichthyosaur the humerus, radius and ulna are the homologues of the same bones in mammals, and at the same time declare that the carpal, metacarpal, and phalangeal bones of the same Ichthyosaur are *not* homologous with bones of the same name in mammals!

Well, some Ichthyosaurs have seven digits and a dwindling eighth. What has become of the sixth, seventh, and eighth digits in mammals? The answer is, that not impossibly they have been suppressed in past geological times, as a *deformity* in the first instance, and afterwards inherited as a *normality*; that is, if mammals ever descended from animals like the Ichthyosaur. Not only have three digits been suppressed, but also a number of phalanges, and the whole carpus, ulna, radius, and humerus have been vastly modified.

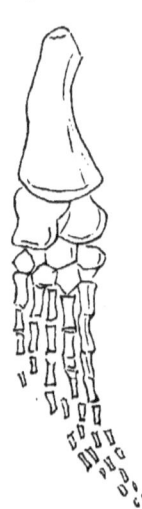

FIG. 93.—Left hand of Plesiosaur, pl. 27, *Book of the Great Sea-Dragons*, by Thomas Hawkins.

[1] Is there any good reason for supposing that Ichthyosaurs were *not* mammals?
[2] See figure of Round-headed Dolphin in Flower's *Osteology of Mammals*, p. 302.

MONSTROSITIES

In this connection it is interesting to note that in the York Museum there is a unique Plesiosaur. It is ticketed *Pl. Zetlandicus* (Phillips) from the Lias of Lofthouse near Radnor. It has a much shorter neck and a much larger head than the ordinary Plesiosaurs. It may be one of the transition forms between the Ichthyosaurs and the Plesiosaurs. Its large head is not unlike that of an Ichthyosaur. It is the only example known of this species.

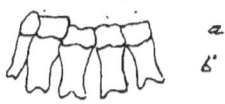

FIG. 94.—(*a*) distal row of carpal bones; (*b*) metacarpal bones of the right manus (dorsal aspect) of Water Tortoise; *Mammals*, by Flower and Lydekker, p. 48.

Then let us compare for a moment the carpal and metacarpal bones of certain animals. In the Water Tortoise we find two regular rows of carpal bones, of five bones in each row, each metacarpal bone articulating with only one of the carpal bones, as shown in Fig. 94; while in the hand of the Plesiosaur (Fig. 93) we find the carpal bones reduced, and huddled up in two irregular rows, the metacarpal bones articulating quite differently from those of the Water Tortoise, although the number of *metacarpal* bones is the *same* in both animals.

Such a difference in the disposition of the carpal bones, and in their articulation with the metacarpal bones, in these two animals may have occurred gradually, but there is no good reason why it could not have occurred suddenly by a process which we would now call *monstrous*, or, to use a scientific term, *teratological*.

There is one other great conclusion to be drawn from the study of the records of sex-digitate men and women. In spite of their marrying individuals in whose families no such abnormality occurred, sex-digitation, as we have seen, persistently appeared among their descendants, so that the notion which some evolutionists hold, viz., that a variation which may suddenly appear becomes soon extinguished by mixing freely with normal types, would seem to be *erroneous*. For sex-digitation, instead of being eliminated by dilution, in many instances becomes *strengthened*. Moreover, the supposition that *one* monstrosity occurring in a species has no chance of survival must be greatly modified when we know that a *whole litter* of puppies can at once appear with stump-tails,[1] and that almost all the puppies of a sex-digitate Bitch were born with a similar feature. It is evident that there would be ample opportunities for these similarly affected animals to *interbreed*, and further fix and perpetuate a monstrosity, even if we did not know that, in some cases, dilution with normal blood will *not* wipe it out.

Many other monstrosities besides sex-digitation have been transmitted to descendants. That quoted by M. Guinard on p. 162[2] is a most curious one. A male Guinea-Pig was kept for a long time in the laboratory of the Veterinary School of Lyons. It had no sign of eyeballs, although the orbital cavities were well formed. After death, it was found that there was no trace of optic nerve, and the foramen for its passage did not exist. Otherwise this anomalous Guinea-Pig was healthy and vigorous, and it lived for a long time. It was kept among normal female Guinea-Pigs, and had numerous children. Among them were found six which, like the father, had no eyeballs! M. Guinard adds:

'M. Desfosses a démontré que cette conformation des organes de la vue est normale chez le Protée.'

Hairlessness of the body is found 'normal' in the hairless race of Chinese Dogs, but it has also occurred anomalously in the Horse, the Ox, and the Dog. Even Birds have been met with which had no feathers. So that we begin to suspect that the hairlessness of

---

[1] I have known of two Bitches, in different countries, which gave a *whole litter* of puppies with stump-tails.   [2] *Précis de Tératologie.*

MONSTROSITIES 303

Man may have occurred *all of a sudden* as a monstrosity, and then may have been inherited and perpetuated.[1] If hairlessness of body happened to carry with it a correlative development of the brain, with superior intelligence, the hairless race would have eventually killed out the hairy ones. And now we get the hairy Ainos, either as reversions, or as *vestiges* of a character which probably was normal in remote times, the Burmese hairy family, and the Russian 'homme chien,' being atavic anomalies.

That the nervous system has a great deal to do with hairiness is shown by the case quoted by M. Guinard (p. 155)[2] of a woman whose trunk became hairy during each pregnancy, and lost its hairy coat in the intermediate periods.

This is analogous to what happens in animals when a change of season and other external conditions, acting on the nerve centres, bring about a change of coat and plumage.

Isidore Geoffroy St. Hilaire considered that the indispensable condition for the transmission of a monstrosity is that the reproductive organs should be intact. Who can tell how many monstrosities which have occurred in geological periods had answered to this condition?

M. Guinard (p. 19)[2] says :—' Ce ne sont pas les anomalies graves qui se transmettent le plus souvent, mais les simples vice de conformation.'

This is only natural, for if the 'anomalies graves' affect the organs of reproduction, there can be no chance of their being transmitted. It is perfectly conceivable that anomalies, which in geological periods could not be inherited, died out with the anomalous individuals. These, for all we know, might have been numerous, and in cases where only *one* fossil anomaly is traceable

---

[1] The Yahgans of Fuegia are hairless, and wear no clothes. They are exposed to the cruelest of climates, and yet do not become extinct. Their extreme sensitiveness to slight warmth is testified to by Darwin. [2] *Op. cit.*

it may possibly have been a monstrosity which was *not* transmissible.

All malformations connected with the extremities of the limbs are hereditary and are easily transmitted from generation to generation. This is a well-established fact, and proved by numerous and well-authenticated examples. But M. Guinard and others have shown that other anomalies, as long as the reproductive organs are efficient, are also frequently inherited, such as hairiness in man, hairlessness in other animals, hornlessness, want of eyeballs, etc. If the anomalous individual cannot procreate, the anomaly must die out with the individual possessor of it.

Fishes lay an enormous number of ova. Supposing that some external influence were adequate to modify the development of *one* fish-embryo, it would be undeniable that a *large number* of ova might be *simultaneously and similarly* modified.

We have known an anomalous influence to pervade a whole litter of puppies; and the probability is that an anomaly of structure might occur in a whole litter of fishes, which might mean *of thousands*! Therefore we must begin to look upon monstrosities, both now and in the past, suddenly appearing in a type, from whatever cause, as *possible factors in the origin of species*.[1]

[1] In order to impress this important idea on the reader's attention, I will quote a parallel case from the Vegetable Kingdom. Among market gardeners there is a *Myosotis* (Forget-me-not) with almost every flower having many more petals than the typical *Myosotis*, which has only five petals. Most of the flowers have eight or nine petals, and many have ten, which is exactly *double* the typical number. The sepals and stamens correspond with the number of petals, so that there can be no doubt that the ten-petalled flowers are *fusions* of two whole flowers, the fusion occurring while they are in their embryonic stage. The *oval* centre of many such flowers plainly indicates a fusion of two. It should be understood that in this form of *Myosotis* it is only the individual flowers which are monstrous. I have ascertained that the ovules contained in the nucules round the base of the conical style, are, many of them, fertilised, and Professor Henslow has it from the grower, Mr. Appleton of Longford, that 80 per cent. of its seeds *reproduce* this particular monstrosity, although insects must frequently cross-fertilise it with pollen of the typical form, which now and again appears on this monstrous plant, and which also may

MONSTROSITIES

No evolutionist can doubt that an anomaly or monstrosity—
(*a*) if it be viable ; (*b*) if it has its reproductive organs in a sufficiently
fit state ; (*c*) if it be inherited—*may* lay the foundation of a new
species, new genus, new order.

That anomalies and monsters are sometimes viable, that sometimes they have their reproductive organs in a fit state, and that
their features are frequently inherited—are sufficiently shown by the
facts quoted.

Whether a new species or genus in its initial stage can hold its

be grown in the vicinity, the Forget-me-not being a very popular plant. The origin of this particular monstrosity is not known. What we have to note is that the *whole plant* is pervaded by this monstrous influence, and that it is *reproduced* through the seed.

There are two other varieties of monstrous *Myosotis* which have come under my notice. That called 'The Jewel,' which was issued by Messrs. Carter and Son in 1893. I have only seen a picture of it. Not only is each flower a *fusion* of two flowers, but the flowers are gathered into a compact head (*capitulum*) somewhat like that of a Scabious, or the double Corn-flower of gardeners. This additional monstrosity is declared to come true from seed.

Then in the *Gardeners' Chronicle* of 8th August 1891, p. 159, to which my attention has been called by Professor Henslow, there is a third monstrous form of *Myosotis*, called *Victoria*. It is a sort of 'hen and chicken' form, all the chickens have flowers which are fusions of two entire individuals of the typical five-petalled form, and the hen in the centre appears to be a fusion of about five distinct flowers. Messrs. Ernst Benary of Erfurt sent it out in 1886, and it is presumed that it sprang from *M. alpestris robusta grandiflora* (Eliza Fonrobert). It comes true from seed. But the curious fact connected with this particular form is this : It appeared in the garden of Mr. W. Marshall at Bexley, who states that he did not receive seeds from either Germany or anywhere else. It is hardy in severe winters. The stem is as thick as a Swan quill, compressed and hollow in the interior, characters which are indicative of a fusion (fasciation) of many separate stems. The drawings in the *Gardeners' Chronicle* are taken from a specimen exhibited by Mr. Marshall, and the editor says :—' It is reproduced from seed, and is now known to have occurred during several years, and possibly may have originated spontaneously in a garden at Bexley, as well as in Germany.'

It may perhaps be questionable whether in this monstrous *Myosotis* the ten-petalled flowers are a fusion of two *separate* flowers while in their embryonic stage. We know that *one* ovum in certain animals is capable of developing twins, and possibly the ovule of a plant may sometimes do so likewise. I consider the ovule as a bud, and the flower as the development of a bud ; so that I can see no objection to the idea that a bud could under certain circumstances develop twins *parthenogenetically*. This hypothesis might account for the flower-buds of this *Myosotis* developing into *fused twins*.

U

ground in the battle of life is a totally different question. This may depend on conditions of the environment. Among fishes we have seen that there are great facilities for the sperm cell to reach the germ cell, and that large numbers of individuals might possibly inherit an anomaly in one generation. This would make the result of the struggle for existence quite different from what it would be if only *one* in a generation inherited the anomaly.

There is no good reason why the species or genus, originating in a monstrosity, should not in various cases be better able to struggle for life than the type of its progenitors. It would all depend on the means of offence and defence, or the means of obtaining food and of resisting adverse influences.

Palæontologists declare that in past ages it was a common phenomenon for a new type to appear, to increase to a maximum, and then gradually disappear, to be replaced gradually by a new type, which in turn followed the same course, to be replaced by a subsequent and more vigorous type. The appearance of a viable and inheritable monstrosity that was better equipped for the struggle of life may possibly account to some extent for this succession of type after type, to go through the same process of increase, decline, and final extinction.

Man, with his monstrous intelligence (not unfrequently with his monstrous foolishness) has killed out and is killing out a number of types, and is replacing them with his own type. What will be the upshot of all this process remains to be seen.

Professor Louis Agassiz wrote:[1] 'However long and frequent the breaks in the geological series may be, in which they would fain bury their transition types, there are many points in the succession where the connection is perfectly distinct and unbroken, and it is just at these points that *new organic groups are introduced without*

[1] Footnote to p. 188 of Dr. Charles F. Holder's *L. Agassiz—Life and Works*.

MONSTROSITIES 307

intermediate forms to link them with the preceding ones.' (Italics are mine.)

Let us take for instance the case of only one character—that of horns. Horned animals may have originated in the first instance as a monstrosity. We find that in the Reindeer both male and female are furnished with horns. In the Eland (*Oreas Canna*) Mr. Selous says the horns of a female were longer than those of a male with the longest horns,[1] but usually in other ruminants, if horned, the female has shorter and more feeble horns. Then we have species in which the female has no horns at all, and finally we have a race of domestic cattle, arising suddenly as a monstrosity, in which both male and female are hornless, and so are Camels, the Musk Deer, and some others. We have no idea how in these animals the suppression of horns may have come about, if their ancestors ever had them.

The appearance of new types in geological times *without intermediate forms* must have been a very puzzling feature in the study of evolution, both to Agassiz and to ordinary Darwinians. Can it be possible that the sudden appearance of monstrosities, which did *not become extinguished* by crossing with typical forms, may have had something to do with these puzzling phenomena?

Professor Alleyne Nicholson, in the Swiney Lectures of 1893, told us that both in the Old and New World, in the Old Tertiary beds, a large number of quadrupeds appear *all at once without any predecessors*, that is, ancestors from which according to the doctrine of evolution these new quadrupeds could be reasonably considered to be descendants.

This is not all. For accompanying this new inroad of unaccountable animals there was an inroad of plants. Between the

[1] 'The longest pair of Eland Bull horns I have seen measured 2 ft. 6 in. in length, the longest pair of Cow horns 2 ft. 10 in.'—*A Hunter's Wanderings in Africa*, p. 206.

upper and lower chalk, the Cycads, which previously were very abundant, now become less and less, and higher groups of plants, all highly specialised, take their places.

In this connection Sir J. William Dawson[1] says: 'We have a great and sudden inswarming of the higher plants of modern types at the close of the lower Cretaceous. In relation to this, Saporta, one of the most enthusiastic of evolutionists, is struck by this phenomenon of the sudden appearance of so many forms, and some of them the most highly differentiated of dicotyledonous plants. The early stages of their evolution may, he thinks, have been obscure and as yet unobserved, or they may have taken place in some separate region, or mother-country as yet undiscovered, or they may have been produced by a rapid and unusual multiplication of flower-hunting insects! or it is even conceivable that the apparently sudden elevation of plants may have been due to causes still unknown. This last seems, indeed, the only certain inference in the case, since, as Saporta proceeds to say in conclusion: "Whatever hypothesis one may prefer, the fact of the rapid multiplication of dicotyledons, and of their simultaneous appearance in a great number of places in the Northern Hemisphere at the beginning of the Cenomanian Epoch, cannot be disputed."'

Quoting from Dr. Newberry,[2] Sir J. W. Dawson at p. 205 says: 'But the most surprising discovery yet made is that of a number of quite large helianthoid flowers, which I have called *Palæanthus*.' It seems probable that the compositæ formed a part of the Cretaceous Flora. Dr. Newberry adds: 'No composite flowers have before been found in the fossil state, and as these are among the most complex and specialised forms of florescence, it has been

[1] *Geological History of Plants*, p. 193.
[2] *Bulletin of the Torrey Botanical Club*, March 1886.

supposed that they belonged only to the recent epoch, where they were the result of a long series of formative changes.'

Now I have always thought that composite flowers may have originated from some ancestral monstrosity—a disc-like fascication of some plant of the nature of *Pimelea, Ixora, Bouvardia Valeriana, Centranthus*, and others,[1] and here is a helianthoid mass of flowers appearing all of a sudden in the Cretaceous flora—without any apparent predecessors to explain its *gradual* evolution! May it not have been the development of a monstrous form appearing suddenly like the composite-like *Myosotis* (The Jewel) mentioned before, which came out of the ordinary *Myosotis* with a scorpioid inflorescence?

Professor Nicholson said that the most rational way to account for this new creation, so to speak, is to suppose that in the Pacific Ocean there was a continent which sank, slowly it may be, and is now no longer traceable, so that the animals which had been there evolved had time to migrate in two directions, to what we now call the Old and the New Worlds. *Their* predecessors are now under the sea, and therefore this theory, unless that continent should come up again, is not likely to be confirmed.

These new quadrupeds, although specialised, had still a generalised structure, and were all five-digited in hand and foot. In the Early Tertiary, as a rule, they had that number of digits, and then a diminution in their number commenced, such as in the Hippopotamus, the Rhinoceros, the Ruminants, and finally the Horse, which has only one digit. Concurrently with the reduction of digits there occurred a reduction in the number of teeth from forty-four to thirty-two in Man, the latter being now reduced to twenty-eight.

There was another feature connected with these new quad-

[1] The reader should note that the disposition of the florets of the disc of several composite flowers is *spiral*, as it often is on an ordinary raceme.

rupeds. In the Early Tertiary both the fangs and crowns of their teeth were short, and as they progressed their teeth became longer, so that their lives became longer also, as their teeth did not wear away so quickly. Moreover, the Early Tertiary quadrupeds, although of large size, had small brains, and this condition was also conducive to a short life, for in the struggle for existence the animal with a large brain has always been able somehow to circumvent the one with a small brain. The early types had no horns.

There seems to me two weak points regarding this theory of a sunken continent in the Pacific Ocean.

(a) If highly specialised types existed on this Pacific continent, what prevented their migrating *before* that particular time? Was it the *isolation* of the continent? If so, when it began to sink it would become *more* isolated! Of course it may be said that the continent was not wholly isolated, but in some place connected with other continents by a neck of dry land such as that which connects Africa with Asia : and that when a portion of this lost continent began to sink, that is, when the grazing or feeding grounds of those animals began to be swallowed up by the sea, and their numbers continuing undiminished, hunger forced them either to die or to migrate by some land communication. There is, however, another and a weaker point in this theory.

(b) Concurrently with the appearance of a new type of quadrupeds in the New and Old Worlds there appeared a new type of *plants*. From the previous Cycads, which are rather fern-like, we pass, without any local predecessors, to plants, which are more or less like those we have *now*, that is, flowering plants.

We now know that a large number of seeds of flowering plants can be transported long distances by strong winds and cyclones, and even on the mud that sticks to the feet of water-birds and waders, and in some cases on their feathers also.

MONSTROSITIES

If these flowering plants existed on this lost continent, one does not quite understand how their seeds could have been prevented from getting disseminated to both the Old and New World *before* this supposed migration took place. One can understand that a certain number of seeds went over with the animals attached to their hair, but what could have prevented the migration of seeds *before* the migration of animals from this lost continent? This is the part of the theory that would seem to require some explanation.

Then Dr. Alfred Russel Wallace seems to insist that no great changes in the continental and oceanic configuration have occurred.

There is another point of importance in this discussion. ' It is a remarkable phenomenon,' says Sir J. William Dawson,[1] ' in the history of genera of plants in the Mesozoic and Tertiary, that the older genera appear at once in a great number of specific types, which become reduced as well as limited in range down to the modern. This is, no doubt, connected with the greater differentiation of local conditions in the modern ; but it indicates also a law of rapid multiplication of species in the early life of genera.'

What is the cause of this *rapid multiplication of species in the early life of genera*?

From what we know of heredity, it is a great controller of form, and therefore in those days there must have been some *efficient disturber of heredity*. Teratogeny may yet throw light on this puzzling question. Hybrids between two genera are certainly not unknown in modern horticulture, and hybrids between two species are common enough ; and it is not impossible that a large proportion of this puzzle may be due to *words*. The older botanists may have given generic names to forms which were no more distinct, *physiologically*, than what we now call varieties. It seems to me that the theory of a sunken Pacific continent in no

[1] P. 222, *op. cit.*

way explains the 'remarkable phenomenon' of 'rapid multiplication of species in the early life of genera.' It is not impossible that in those days what we would call species and genera may have been more readily hybridisable, and by steady change of conditions and consequent physiological differentiation, up to modern times, the process of *natural* hybridisation may have become more difficult and rare.

Professor Huxley[1] says : 'In former periods of the world's history there were animals which overstepped the bounds of existing groups and tended to merge them into larger assemblages. They show that animal organisation is more flexible than our knowledge of recent forms might have led us to believe ; and that many structural permutations and combinations of which the present world gives us no indication may nevertheless have existed.'

This seems very natural, for the further we go back in time the *less inheritance* the species may have had at its back, and therefore the less time there may have been for the creation of a *fixed habit of the nerve-centres* in animals, which regulate everything. Consequently they would be more variable. On the other hand, the further we go forward in time, the more the habit of inheritance would become confirmed, and therefore the more *fixity of characters* would ensue. The *oldness* of the inheritance would then have made it *too strong* to admit of much change, unless man interfered with the natural process by artificial selection.

Professor Huxley (p. 89) warns us not to be in a hurry to conclude that because no organic remains are found in a deposit, that therefore 'animals or plants did not exist at the time it was formed,' for he has known fossil remains *quite dissolved away*, leaving no trace but the casts which formerly surrounded them. It might so happen that even the casts might be destroyed.

[1] *Lectures on Evolution*—(Science and Hebrew Tradition), p. 102.

The unfortunate thing is that in all these questions we try to evolve theories with an insufficient number of facts to build upon. For it cannot be said that either the geology or the palæontology of our earth is *all* known.

We do not know yet whether a teratological form simulating a totally different type from that of its immediate progenitor is possible. All our present knowledge tends to show that heredity would restrict a great deviation from the normal type.

Darwin and his followers, in order to get over this difficulty, supposed that there was no such thing, now or ever, as *identical* heredity, and that every new being *varied* in some way from its progenitors. Therefore there can only be heredity *with variation*, as we see it among domestic animals. And as these variations may be very small, the great differences we see in animals is accounted for by the *accumulation* of small differences, whenever useful, over long periods of time, through natural selection. There may have been many intermediate forms between the present types, but these have become extinct in the struggle for existence, or from other causes.

M. Guinard[1] says: 'Il est impossible de séparer variation et anomalie, car en sens vrai du mot, la variation est toujours une anomalie.'

I would add that it is impossible to separate either from *monstrosities*. For if we are to classify as simple variations the absence of the tail in a Dog, or of horns in an Ox, why not other monstrosities?

'Pour ces raisons,' he continues, 'nous ne séparerons pas, dans notre étude, la variation de l'anomalie, qui sont l'une et l'autre des déviations du type spécifique.'

On p. 7, however, he seems to put aside all this logic; for he

[1] *Précis de Tératologie*, p. 5.

says : 'A case of polydactylism in the Horse, or of double uterus in a Woman, or of prolongation of the tail in Man, must be considered as a simple return to an ancestral type, that, is *reversions*—anomalies, as far as the individual case is concerned, but not concerning the *species* to which it belongs (!) '

One would like to ask—Is there any law which limits reversions? If a polydactyl foot in a Horse is a reversion to the pentadactyl foot of some ancestral form, why is not sex-digitation in Man a reversion to something like the octodactyl hand of the Ichthyosaur?

M. Guinard appears to me to lay too much stress on what he calls *specific type*, as if there were any truth in the antiquated dogma that specific types are unchangeable and are not modifications of some other pre-existing types. If supernumerary mammæ in a Woman, and horns in a Horse, are reversions to conditions existing in *other* species, it appears to me that there is no good reason for supposing that reversions can go, not only beyond the species, but beyond the genus, and the order too, for it seems to me that artificial subdivisions in the flow of what we call 'vital energy' are as *artificial* as a classification of the different portions of a waterfall would be !

Dr. Wallace, in *Darwinism*, p. 101, says : 'On comparing the variations which occur in one generation of domestic animals with those which we know to occur in wild animals, we find no evidence of greater individual variation in the former than in the latter.'

I suppose by variation he means the ordinary small variations which are peculiar to *every individual*; for on p. 100 he refers to monstrosities only with reference to artificial accumulations of something beautifully novel, strange, or amusing. Some of these artificial monstrosities would be injurious, such as the tumbling of Pigeons. In his work on *Darwinism* I cannot find that Dr. Wallace even suspected that a new type of animal or plant *might be founded* on a monstrosity appearing *suddenly*, and in another place he rejects this idea.[1]

M. Guinard finds that he cannot make any distinction between a variation and an anomaly, and I find myself unable to make any distinction between variation, anomaly, and *monstrosity*. They are only degrees of the results initiated congenitally during embryonic development, under different conditions of physical surrounding. The latter may give rise to a new type *suddenly*, provided it is viable and reproducible; the former may give rise to a new type by *gradual* accumulation of a variation in a certain direction. It seems to me that we must have them *both* as factors in the *origin of species*.

Dr. Wallace's own observations of certain facts in nature can only be explained by initiation as monstrosities, and not by gradual accumulation of minute variations under adaptive conditions. In his *Travels on the Amazon*, p. 58, he says: 'What birds can have their bills more peculiarly formed than the Ibis, the Spoonbill, and the Heron? Yet they may be seen, side by side, picking up the same food from the shallow water on the beach; and on opening their stomachs we find the same little crustacea and shell-fish in them. So of the fruit-eating Pigeons, Parrots, Toucans, and Chatterers.'

It has been assumed by some writers on Natural History that every wild fruit is the food of some bird or animal, and that the various forms and structure of their mouths may be necessitated by the peculiar character of the fruits they are to feed on; but there is more of imagination than fact in this statement; the number of wild fruits furnishing food for birds is very limited, and birds of the most varied structure, and of every size, will be found visiting the same tree.

Is not all this clear evidence that it is *not* always the habit of

[1] See *Fortnightly Review* for May 1893, p. 657.

the animal which gradually adapts its structure to surrounding conditions, but that *sometimes* a monstrosity may occur which with an abnormal structure must either adapt its habits to its surroundings or must cease to live? Those monstrosities which *can* succeed in adapting themselves to surrounding conditions live, and reproduce themselves, to startle and puzzle us with the fact that so many strange and abnormal structures are suited to *identical* habits! And so we have Ibises, Spoonbills, and Herons, among Waders, leading the same life; and Pigeons, Parrots, Toucans, and Chatterers, among frugivorous birds, leading also an identical life among trees.

There seems no reason for doubting that a monstrous individual, having found its monstrosity *useful* in the battle of life, would acquire in some cases *habits consonant with the monstrous limb or limbs*, and we might be misled by thinking that the anomalous structures were brought about by the habits gradually modifying the structure, while the exact reverse may, in many cases, have happened, just as an athletic man would take to athletic habits *because* his muscles were originally well developed.

That new habits, to which organisms may have been driven, may be engendered by slow and minute steps of variation in structure may be too true, but this in no way excludes the great probability that many new and monstrous variations, as we would now call them, may have originated *in one great and sudden step*,[1] improving afterwards if the individual were viable and reproducible. That anomalous structures are often supplied with not only bones, but also muscles, vessels, and nerves, as if they were inherited, and ready equipped for use, is sufficiently proved by the supernumerary digits. In many cases, both structurally and functionally, these could not be distinguished from the normal ones.

[1] Professor W. Kitchen Parker thought so also (*Mammalian Descent*, p. 93).

MONSTROSITIES

No one who gives this subject any attention can look over Mr. Stebbing's *History of Crustacea*, and also the specimens in the Natural History Museum, without becoming convinced that many of them must have originated *suddenly* as monstrosities, brought about by some atomic disturbance, during the development of the embryo, and that subsequently natural selection may have further modified the monstrous parts, the animal having found them *useful* in the battle of life, or at all events found that they did *not seriously interfere* with its continued existence.

Look at that unfortunate *Dorippe japonica* of Fig. 10 (Stebbing's *Crustacea*). It descended from ancestors that had *eight* fully developed walking legs, but Dorippe has only *four* walking legs, and the other four are ludicrously dwarfed and *stuck on its back*. In the Natural History Museum there is a specimen ticketed *D. dorsipes*! Then look at the figures on Plate vii. Every one of the specimens has either one pair or two pairs of atrophied legs, just as if you saw an adult Man with a normal body and two little 'wee legs' at the end of it, like those of a newly born baby.

Stebbing says: 'To account for their dorsal position various reasons have been suggested. Herbst says the Crab can run either way up.[1] Another view is that these hind-legs lift foreign objects on to the carapace, and a third, that they help to repel animals that might otherwise tread on the Crab's back.'

Whatever may be the use the Crab makes of them now, the dorsal legs are best explained as having originally been brought about as a *monstrosity*, which the animal had the good sense to make the best use of, if it had to exist at all.

The curious part of it is that, with all its anomaly and suppression of four legs from the function of progression, *Dorippe* has

[1] I suppose, when it gets tired of running one way, it throws itself on its back and runs on the other way, so that it can tire out its enemies!

318   STUDIES IN THE EVOLUTION OF ANIMALS

a wide range, and includes several species. *Dorippe facchino* is found both in the Mediterranean and at Hong-Kong. There is no good reason to suppose that two such monstrosities could *not* have originated *independently* ;[1] although it is conceivable that the ova, in past ages, may have been somehow carried from one place to the other, especially when we know that there is geological evidence that the Mediterranean Sea in past geological times—Palæozoic, Mesozoic, and even Tertiary—communicated with the Indian Ocean across north Arabia and also with the Bay of Bengal across north India.[2]

'Facchino' in Italian means 'porter,' and the specific name may have been given in allusion to the supposition of its using its dorsal legs like arms to carry loads!

A writer in *Chambers's Journal* for November 1890, on 'Jungle Notes in Sumatra,' p. 663, says: 'These Hornbills are very remarkable birds. I cannot imagine any system of natural selection which could have developed those preposterous-looking beaks. Was it because those with the largest beaks could best defend their families against Monkeys and Snakes? But what size of beak did they start with? If they were so persecuted a race, would not their enemies have exterminated them before they had time to develop their weapons? You cannot, I suppose, allow less than five thousand years for the process, and if they had to begin with a beak the size of a Fowl's, the Monkeys alone would "wipe them out" in ten years.'

There are many kinds of Hornbills in existence, commencing from that with a bill not larger than that of a small Toucan, to

[1] Mathematicians might tell you that the chances of two identical forms having originated independently are as 'infinity to one.' But is this so? How many crystals of this system, or of that, acquire the *same* geometrical form, although they originated *independently*, and are made up of different elements?

[2] *Fragments of Earth Lore*, by Professor James Geikie.

that of the Great Hornbill. In some the crest on the upper mandible is small. It is quite conceivable that the crest enlarged gradually up to a certain point, and that by a sudden monstrous enlargement it reached the size we see on the mandible of the Great Hornbill.

Then there is *Toccus elegans* and *Toccus Montieri*,[1] which have enormous bills like those of the Great Hornbill, but without a trace of the monstrous crest. The Condor has a crest on its upper mandible not unlike that of the Hornbills.

Mr. F. W. Headley[2] writes: 'The Hornbills are a puzzle. The extreme shortness of the hand bones, a ridiculous anticlimax following upon so grand an ulna and so portentous a humerus, might suggest that they were once better flyers, and that the wing is slowly undergoing reduction. But the mountainous beak seems to show that colossal bones are an ancient heritage of the family, and that even feeble flight might have been difficult had they not become hollow.'

From this it would seem that Mr. Headley would suggest that the immense beak of the Hornbill was inherited from some ancestor with a colossal body, but this does not touch the *origin* of the immense crest, seemingly useless, on the upper mandible of the Great Hornbill. As these birds are frugivorous, this colossal bill would seem to be rather a nuisance among the branches of trees.

There can be little doubt that in the course of evolution an occasional monstrosity, as a *sudden large variation*, would facilitate the interpretation of such vagaries in nature and of many a difficult problem in evolution.[3]

[1] *Proc. Zool. Soc.* 1865, p. 86.
[2] 'The Air Sacs and Hollow Bones of Birds,' *Natural Science*, No. 21, vol. iii. p. 352, Nov. 7, 1893.
[3] The extinct Pterodactyls had immense bills, not unlike those of the Hornbills, and among *them* we may yet find this ancestry of the massive bill of the Hornbills!

The evolutionary osteologist who may be still wedded to the theory that every type of animal *necessarily* came out of another type by *slow and gradual modifications, accumulated by natural selection*, so as eventually to bring about a marked difference between the two, may still be hoping that future palæontological discoveries may bring to light all the intermediate links. He may obtain more links than he possesses now, just as he possesses graduated links between the various existing Cats and Dogs, Pigeons and Fowls.

But even if the whole of the geological formations of the earth were explored, I should not anticipate that he would find all the links he may hope for. Teratology plainly indicates that 'leaps,' *without* much trace of graduated intermediate links, can occur now, and therefore most probably similar leaps were not uncommon in the days when the physical conditions of life must have been far less equable than we now know them.

It would be curious if the study of abnormalities should give us an insight into the origin of what we call *normalities*.

Mr. J. A. Thomson[1] writes: 'Among back-boned animals we recall the teeth of the Shark, and the sword of the Sword-Fish, the venomous fangs of Serpents, the jaws of Crocodiles, the beaks and talons of Birds, the horns and hoofs of Mammals. Now we do not say that these and a hundred other weapons were from their first appearance weapons, indeed we know that most of them were not. . . . By sheer use a structure not originally a weapon became strong to slay.'

Just so; and the sudden appearance of a monstrous part would produce one of two results:

(*a*) Either it would have no appliances of muscles and nerves to utilise it in the struggle for existence, and then it would be

[1] *Study of Animal Life*, p. 34.

MONSTROSITIES 321

rather an encumbrance and a disadvantage, and its possessors might soon become extinct; or

(*b*) It would have appropriate nerves and muscles; and then its owner would soon learn to make use of it for its own advantage, and being congenital it would be inherited.

We have seen that supernumerary digits may be either simple appendages of the skin, without phalanges, muscles, or nerves; or they may be as well furnished with all three, and indistinguishable from other digits. A savage born with hands and feet like those

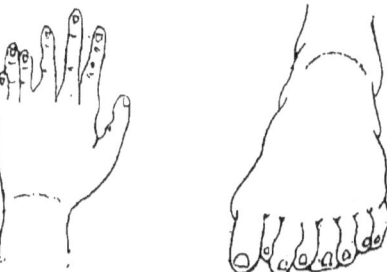

FIG. 95.—Monstrous hand and foot from *Anomalies de l'Organisation*, by Isidore Geoffroy St. Hilaire, vol. iii.

shown in Fig. 95, would indubitably have more power in both than another born with hands and feet like those of Fig. 96, or perhaps than others born with the ordinary number. And so of other monstrosities which may have occurred in geological times, and which may have had their monstrous parts furnished with means for using them as weapons either of offence or defence, or for maintaining their hold on their surroundings.

We have little knowledge as to the process by which some cells of the embryo turn into muscles, others into nerves, and so forth. All we know is that after the egg has subdivided itself into a

x

certain number of cells, it begins to take shape, and layers are formed; some cells forming muscle, others, nerves, and so on, and ultimately completes a creature like its parents, and this likeness we call *heredity*. But we know also that sometimes the regular process is disturbed, and monstrous forms are the result, and tera-

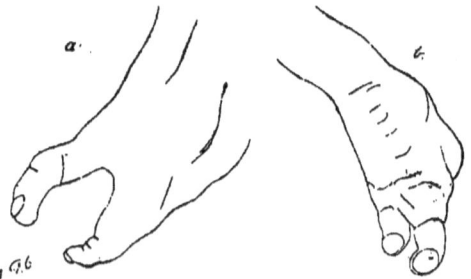

FIG. 96.—(*a*) First and fifth digits of foot; (*b*) Ring and little finger of hand: from *Anatomie Pathologique*, Tom. 2, Livrais. 38, Pl. 1, of Cruvelhier.

togenetic investigations may one day tell us something more definite about the genesis of such monstrosities.

It is curious that most Darwinians infer a *great deal* regarding the *past*, from what has gone on in *domestic* Pigeons, and very few infer *anything* regarding past phenomena of evolution from the *abnormal and monstrous forms* which occur in man and in domestic animals!

To conclude: however reluctant we may be to admit it, I believe we shall have to admit that small and gradual accumulations of variations do not cover the whole phenomena before us, and that occasionally large and extraordinary variations may have occurred in the past, which we would now call monstrosities; and that they may have at times laid the foundation of a new order of progress.

When once a viable individual with a reproducible monstrosity

MONSTROSITIES

occurred, it would almost inevitably follow that more nervous energy would be directed to it from generation to generation by use, *if it could be used*, and it would increase in size until it reached a maximum, beyond which it would become a drawback in the struggle for life. Such a drawback must, I surmise, have happened to the extinct *Elephas Ganesa*, with his enormous ten-feet tusks.[1] It must be plain also that if the tail-coverts of the Peacock were to become twice or thrice their present length, they would become an encumbrance and a disadvantage to the bird. The long nose of the Elephant, the immense horns of the extinct Irish Deer, the sword-shaped upper jaw of the Sword-Fish, the turned down lower jaw tusks of the Dinotherium, and many other similar extraordinary phenomena in animals, may have started as *monstrous features*, all of a sudden, *in one generation*, and may have afterwards become strengthened and larger by use. To be more scientific, we shall call these 'jumps' or 'sports' in the evolution of animals 'teratological variations,' so that the believers in the gradual accumulation of small variations as the *sole* method of evolution may not be shocked.

It has been a wonder to me why Dr. Wallace and other Darwinians ignored or rejected such a palpable help to the doctrine of evolution, as the fact that monstrosities do occur, would afford. This doctrine requires the support of every known fact, whether that of minute accumulated variations, through many generations, or that of a large variation produced all of a sudden. There is evidence to show not only that such large and sudden variations, at one birth, do occur, but that they are reproduced, and are not always obliterated by crossing with normal forms.

If you reject what palpably ought not to be rejected, you

[1] Gordon Cumming records a tusk 20 feet 9 inches long ! See *Royal Natural History*, vol. ii. p. 546.

324   STUDIES IN THE EVOLUTION OF ANIMALS

increase the difficulties of accounting for the whole phenomena of evolution.

In a recent and remarkable inaugural address, at Oxford, by the President of the British Association, Lord Salisbury stated that mathematicians do not grant the time of 'many hundred millions of years' for the transformation of a Jelly-fish into a Man. 'Lord Kelvin limited the period of organic life upon the earth to a hundred million years;' and Professor Tait further cut down the time from a hundred to ten million.

Without admitting that these mathematicians, like the Pope, are *infallible*, and without admitting that any *particular* Jelly-fish was ever converted into a Man, it is evident that if monstrosities, or, in other words, *large* and sudden variations, were admitted as *possible factors* in the creation of species, by the method of evolution, the time needed for the transformation of the lower into the higher organisms would be *much curtailed*. A writer in the *Saturday Review* of 11th August 1894, commenting on Lord Salisbury's address, says: 'But are the demands of the two parties (physicists and biologists) necessarily as incompatible as Lord Salisbury seems to suppose? If the passage from the condition of a protozoon to that of a vertebrate, in the case of an individual, can be accomplished in a very few months, is it so certain that similar changes have always been so extremely slow in the history of a race?'

GENERAL CONCLUSIONS

'I am convinced that any student of the subject who will cast aside his books—supposing that they have not already bred a habit in his mind of seeing only "in accordance with verbal statement"—and go directly to nature, to note the actions of animals for himself—actions which, in many cases, appear to lose all significance when set down in writing—the result of such independent investigation will be a conviction that conscious sexual selection on the part of the female is not the cause of music and dancing performances in birds, nor of the brighter colours and ornaments that distinguish the male.'

The Naturalist in La Plata, by W. H. HUDSON, p. 287.

GENERAL CONCLUSIONS

IF the reader has been kind enough to read carefully through the foregoing pages, he will be, I feel sure, bewildered with the mass of diction that I have placed before him, and with the numerous references that it has entailed.

To endeavour to quiet his brain, and relieve him of that chaos of impressions which I have been trying to make on his convolutions, and in order to convince him of the truth of what I said, as I am myself convinced, I will gather up all the threads of my discourses into some *General Conclusions*.

From the evidences before me it has become my belief that:

(*a*) The rosettes of the Jaguar and Leopard are pigment-pictures of ancestral calcareous armour, which consisted of bone-rosettes; and that the markings of these Cats are the closest resemblance we have left to this sort of inheritance. In other mammals modifications of ancestral rosettes have been very great, so that the vestigial rosetting is now in many cases hardly recognisable;

(*b*) In the Jaguar and Leopard, these original picture-plates have been *maintained* by natural selection, while in others, their disappearance has been brought about, more or less, also by natural selection;

(*c*) Among the ungulates, the dappled Horse, the Zebu, and the Giraffe are the nearest resemblances we now have to the original ancestral picture-plating, and that all other markings, whether spotting or striping, as in the Cats, are modifications of dappling or rosetting;

(d) Many mammals which have lost all vestige of rosetting have still retained vestiges of ancestral carapacing, that is, a *picture*-vestige of dorsal *armour* and ventral *unarmour*, such as still exist in Armadillos and Pangolins, and that the contrasts of colour between the dorsal and ventral surfaces mean that the ventral surface lost its armour *earlier* than the dorsal surface, and therefore its hair had time to differ in coloration from the dorsal surface which lost its armour at a later period ;

(e) The contrasts of colour that we see so frequently in the muzzle, round the eyes, on the hands and feet, of so many mammals mean the same thing, viz., that those parts lost their armour before the rest of the body, and had time to differ in coloration, after the hair replaced the armour of those parts. Time, of course, has often obliterated all these contrasts of coloration, and in many cases has substituted a uniform self-colouring ;

(f) The Rhinoceros has passed through similar stages. The *Rhinoceros Sondaicus* carries still the plates of its Armadilloid ancestor, while the Indian Rhinoceros has vestiges of ancestral calcareous plating in its hide-knobs. In some cases the Indian Rhinoceros is only partially knobbed, but Sir J. W. Flower[1] gives one of this species *knobbed all over*, taken from a photograph of an animal living in the Zoological Society's Gardens ;

(g) Judging from the very numerous mammals that are either rosetted, or spotted, or striped, or have ringed tails, or have contrasts of colour between the upper and lower surfaces, it seems probable that *all* existing mammals, including marsupials, descended from armoured ancestors ;

(h) The callosities on the legs of the *Equidæ*, and of some ruminants, seem to be vestiges of leg glands secreting some odoriferous substance which, smeared on long grass as the animals

[1] *The Horse*, Fig. 9.

moved about, enabled them to track their associates; certain other ruminants have lost these skin glands, and, instead, have vestiges of them, shown by tufts of hair contrasted in colour and length with the surrounding surface;

(*i*) There is suspicion that the one big digit of the hand and foot of the Horse was in origin *double*, like that of the ruminants. In the latter, the two metacarpal and metatarsal bones fused into a 'cannon' bone, the phalangeal bones remaining *separate*; while in the Horse the metacarpal and metatarsal, *and* the phalangeal double bones, fused into single bones throughout the series. There is some evidence to show that the Horse is more closely related to the ruminants than he is to the Rhinoceros;[1]

(*j*) As monstrosities occur now, they may, with greater reason, have occurred in past ages. If this be admitted, it will be seen that a much shorter time can be allowed for bringing about great modifications in the species of the present and the past times than would be needed by slow and gradual accumulation of minute variations, which is the theory accepted by most biologists.[2]

[1] Since writing these conclusions, I came across Mr. Bateson's *Materials for the Study of Variations*. In Fig. 120 he gives a 'solid-hoofed' Pig, with the iii. and iv. digits fused into *one* large digit throughout, like that of the Horse.

[2] *Mesoplodon Layardi* is a beaked whale, with only two strap-like teeth. Professor Moseley, of the *Challenger*, observes that 'these two teeth in the adult animal become lengthened by continuous growth of the fangs into long curved tusks. These arch over the upper jaw or beak, and, crossing one another above it at their tips, form a ring round it and lock the lower jaw, so that the animal can only open its mouth for a very short distance. . . . How the animal manages to feed itself under these circumstances is a mystery. . . . That these enormous teeth can be of no possible advantage to their owner appears perfectly clear; and they must probably be regarded as affording an instance of semi-monstrous development analogous to the one displayed by the tusks of the Babirusa.'
—*Roy. Nat. Hist.*, vol. iii. p. 35.

APPENDICES

'It adds very much to the interest with which we regard them, if, by studying the general causes to which they are due, we can explain their origin, and thus to some extent understand the story they have to tell us, and the history they record.'
Beauties of Nature, by SIR J. LUBBOCK
(Configuration of Valleys).

APPENDIX A

IN addition to those mammals which are given in the preceding pages, the following outline drawings of badly mounted specimens in the Natural History Museum will give the general reader some idea of the number of widely differing animals which are either rosetted, spotted, blotched, or striped in various fashions. Some have only a *vestige* of ancestral markings.

The names and colours are taken from the Natural History Museum. The colours from age are likely to become faint.

334 STUDIES IN THE EVOLUTION OF ANIMALS

No. 1. Hybrid cub of Lion and Tigress. Light fawn, striped brown.

No. 2.—Ocelot (*Felis pardalis*). Light fawn, with brownish rosettes.

APPENDICES

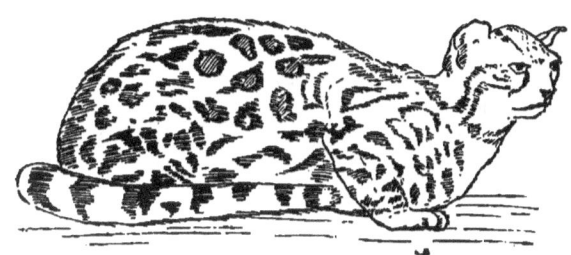

No. 3.—*Felis macroura.* Fawn, with black rosettes and brown centres.

No. 4.—Marbled Cat (*Felis marmorata*). Light brown, spotted and streaked black.

336 STUDIES IN THE EVOLUTION OF ANIMALS

No. 5.—Norway Lynx (*Lyncus lupulinus*). Light fawn, darker on back with black spots.

No. 6.—Civet (*Viverra civetta*). Dirty white, spotted and striped black.

APPENDICES

No. 7.—Tasmanian Wolf (*Thylacinus cynocephalus*). Greyish fawn, banded brown.

No. 8.—*Paradoxurus Derbyanus*. Fawn, banded dark brown.

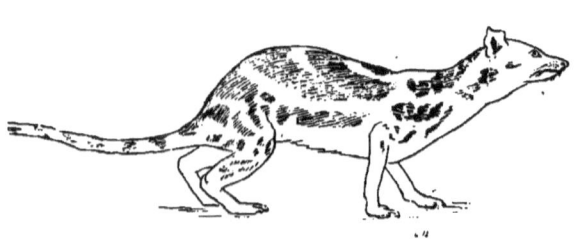

No. 9.—Delundung (*Prionodon gracilis*). Dirty white, piebald and spotted dark brown.

No. 10.—Nepal Linsang (*Prionodon pardicolor*). Dirty white, spotted brown.

APPENDICES 339

No. 11.—Tigrine Genet (*Genetta tigrina*). Light fawn, spotted with dark brown.

No. 12.—Richardson's Genet (*Volana Richardsoni*). Light fawn, spotted dark brown.

340 STUDIES IN THE EVOLUTION OF ANIMALS

No. 13.—*Galidictes vittata*. Grey, striped brown, with dots between the stripes.

No. 14.—Mascarene Striped Mungoose (*Galidictes striata*). Dark brown, with dirty white stripes.

APPENDICES 341

No. 15.—South African Striped Mungoose (*Crossarchus fasciatus*). Grey, banded black.

No. 16.—Paca (*Cælogenys paca*). Dark brown, spotted with white in longitudinal rows.

342 STUDIES IN THE EVOLUTION OF ANIMALS

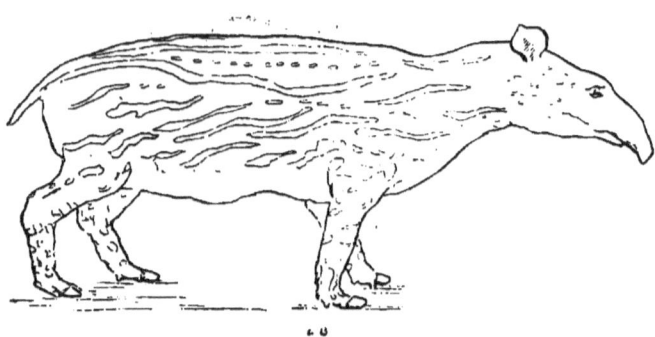

No. 17.—Common American Tapir (*Tapirus terrestris*). Brown, striped and spotted white.

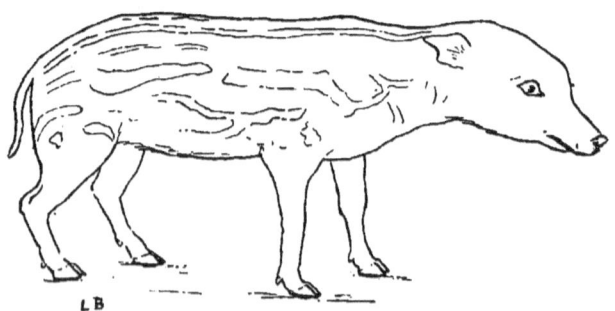

No. 18.—*Sus Andamanensis* (young). Brown stripes alternating with dirty white stripes.

APPENDICES 343

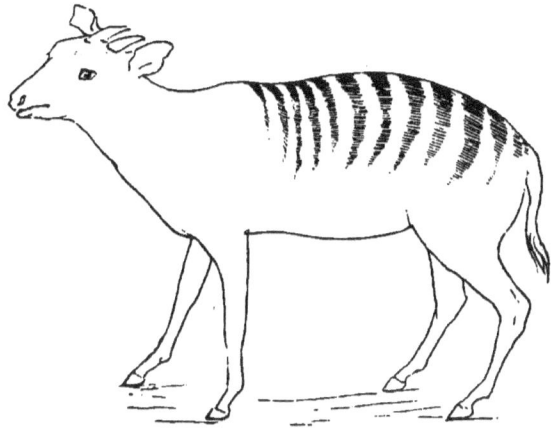

No. 19.—Banded Bushbuck (*Cephalolophus doriæ*). Deep fawn, banded black.

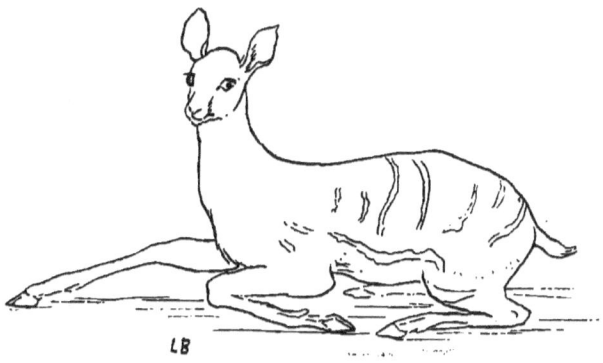

No. 20.—Banded Bushbuck (*Tragelaphus scriptus*), young. Fawn, striped and spotted white.

344 STUDIES IN THE EVOLUTION OF ANIMALS

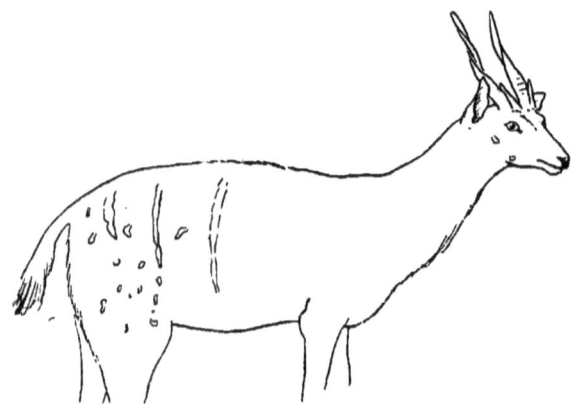

No. 21.—African Bushbuck (*Tragelaphus scriptus*). Grey-fawn, striped and spotted white.

No. 22.—Broad-horned Antelope (*Tragelaphus euryceros*). Yellowish brown, striped white.

APPENDICES

No. 23.—Inyala Antelope (*Tragelaphus Angassii*). Fawn-grey, striped and spotted white.

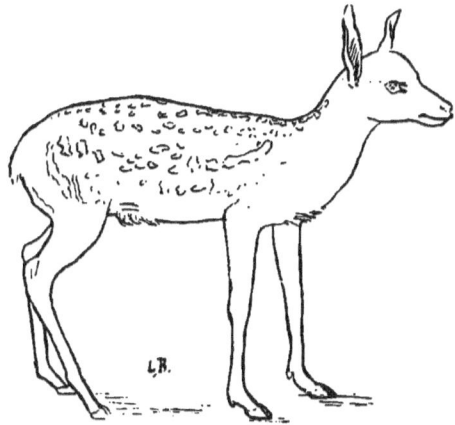

No. 24.—Barbary Deer (*Cervus barbarus*), young. Deep fawn, with white spots.

346   STUDIES IN THE EVOLUTION OF ANIMALS

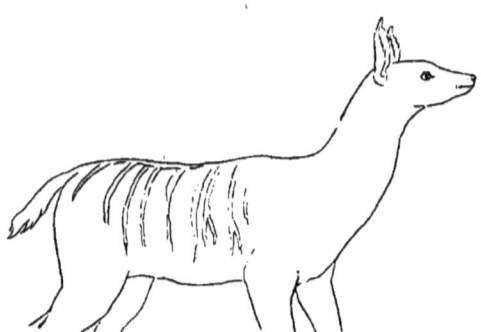

No. 25.—Young of Kudu (*Strepsiceros kudu*). Striped white.

No. 26.—Fallow Deer (*Dama vulgaris*). Deep fawn, spotted white.

APPENDICES 347

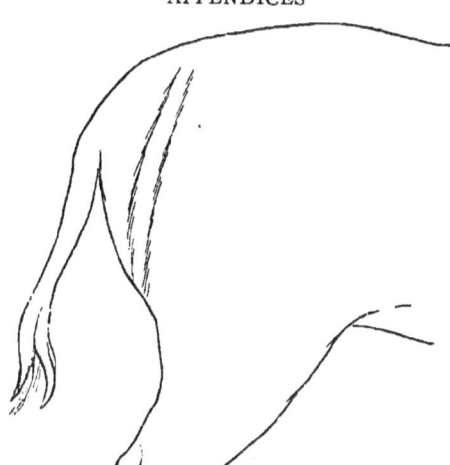

No. 27.—Hindquarters of Water Buck (*Kobus ellipsiprymnus*). Dark grey, with white stripe on rump.

No. 27A.—Mouse Deer (*Tragulus Meminna*). Brown, striped and spotted white.

348   STUDIES IN THE EVOLUTION OF ANIMALS

No. 28.—Lord Derby's Chevrotain (*Tragulus Stanleyanus*) and Kanchil (*Trag. Javanicus*).
Both very young, and having only *vestiges* of stripes.

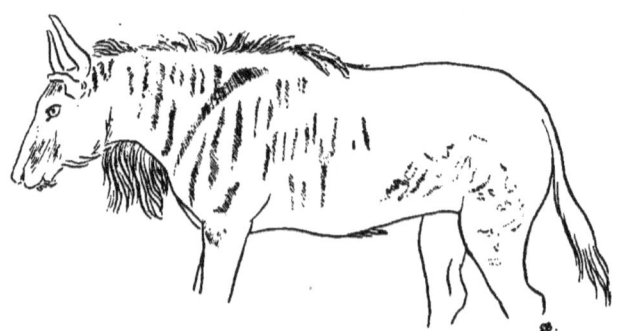

No. 29.—Brindled Gnu (*Comochetes gorgon*). Roan, striped with brown.

APPENDICES 349

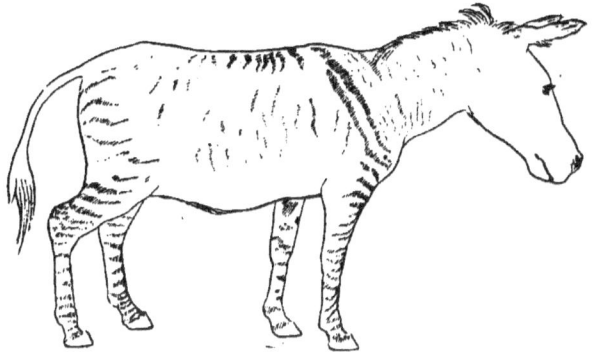

No. 30.—Hybrid between *Equus Zebra* and *E. Asinus*.

No. 31.—Hybrid between *Equus Zebra* and *E. Hemionus*.

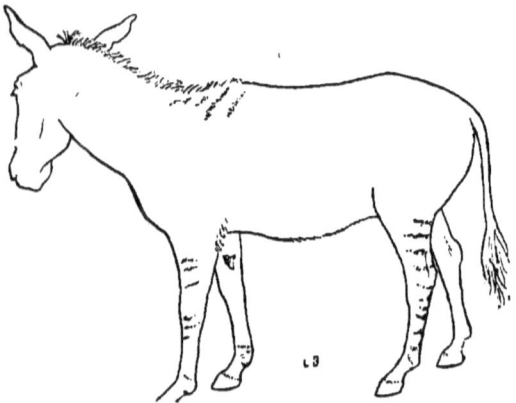

No. 32.—Hybrid between *Equus Asinus* and *E. Hemionus*.

APPENDIX B

IN addition to the Cats mentioned in Part I., the following are also from Mr. Elliot's monograph of the *Felidæ*. They will serve to emphasise the fact that stripes originate in a fusion of spots.

No. 1.—Bushy tailed red spotted Cat (*Felis Euptilura*). Grey, spotted orange-brown; chest transversely barred; the spots are like solid distorted rosettes; the abdomen is pale, and the tail is ringed.

No. 2.—Bokhara Chaus (*F. Caudata*). Bluish grey, trimmed with yellow, and spotted with black; abdomen pale, and tail ringed.

No. 3.—Siberian Lynx (*F. Cervaria*). Grey, spotted with black; abdomen pale, and tail half ringed and half black.

No. 4.—Spanish Lynx (*F. Pardina*). Leopard colour, spotted with black; the abdomen is pale.

No. 5.—Painted Cat (*F. Scripta*). Olive-grey with black distorted rosettes; abdomen pale, and tail ringed.

No. 6.—Little red spotted Cat (*F. Javensis*). Leopard colour or grey, spotted and barred brown; abdomen pale, and tail ringed or spotted.

No. 7.—Shaw's Cat (*F. Shawiana*). Of a dark Leopard colour, spotted with black; abdomen pale, and tail ringed.

No. 8.—Bay Lynx, or American wild Cat (*F. Rufa*). Bluish-grey, rosetted and spotted with black; abdomen pale, and the short tail partly ringed.

No. 9.—European wild Cat (*F. Catus*). Olive-brown, trimmed with yellow and striped black, Tiger-fashion; tail ringed.

No. 10.—Molina's Guiana Cat (*F. Colocolo*). Whitish-grey, broadly striped brown; abdomen pale, and tail ringed.

No. 11.—Pallas's Tibetan Cat (*F. Manul*). Yellowish-grey; loins, haunches, and fore-legs barred; hands and feet yellow; tail ringed.

No. 12.—Chaus (*F. Chaus*). Brownish-grey, almost plain, with transverse stripes on chest and legs; abdomen yellow, and tail ringed.

No. 13.—Caracal (*F. Caracal*). Rich brown, with only vestigial striping in parts; abdomen pale.

No. 14.—Little Malayan red Cat (*F. Planiceps*). Greyish-brown, with vestigial stripes; spotted on the abdomen, which is pale, and barred on the legs.

No. 15.—Northern Lynx (*F. Canadensis*). Grey, with vestigial stripes; its short tail is tipped black.

No. 16.—Bornean red Cat (*F. Badia*). Yellowish-brown, with vestigial stripes on chest.

No. 17.—Eyra (*F. Eyra*). Yellowish-brown, wholly *plain*; abdomen pale.

No. 18.—European red Lynx (*F. Lynx*). Reddish-brown on back and legs; abdomen pale.

No. 19.—Yaguarundi (*F. Yaguarundi*). Like the Puma, but with stripes on the abdomen, and spots over the eyes like those of the black-and-tan Dog.

N.B.—As there are no two Cats exactly alike, although of the same species, two authors describing the same species from different specimens will describe it somewhat differently. The variations in the same species and the transitions from species to species are simply *innumerable*.

Better than any description in words will be found the pictures of Cats in vol. i. *Royal Natural History*, and those of Mr. Elliot's *Felidæ*, the mounted animals in the Natural History Museum, and the live animals in the Zoological Society's Gardens.

APPENDIX C

IN addition to the Horses given in Part II., the following short descriptions, taken from specimens in the streets, will give some idea of the innumerable variations of marking which are to be found among these animals. These are some of the most striking deviations from the ordinary colorations one sees everywhere.

No. 1.—A light bay Horse, brindled on the neck with black, like the striping of a Cat.

No. 2.—A bay Horse on his back and neck had distinct faint stripes like those of the Zebra.

No. 3.—An omnibus Horse, of a strawberry-roan colour, was faintly striped on his flank, and spotted faintly on his hindquarters.

No. 4.—A dappled grey Horse with three transverse stripes on upper part of fore-leg.

No. 5.—It is no uncommon thing to see the mane of dark-grey Horses barred white and black like that of the Zebra.

No. 6.—Bay omnibus Horse with white spots on his flank like those of the spotted Deer.

No. 7.—A dun-coloured Horse, of the colour of a bath-sponge, with cream-coloured dapples.

No. 8.—A flea-bitten Horse is sometimes dappled on his haunches.

No. 9.—A sponge-coloured Horse with a patch of black reticulations in front of his tail-root and a suffused sootiness along the ridge of his neck.

No. 10.—A roan Horse with rows of about five spots on his flank inclined in the direction of his ribs.

No. 11.—A Pony partly clipped—the unclipped part was almost white, while the clipped part was dark-grey, dappled white.

No. 12.—A dark-grey Horse dappled all over with pale tan or dun colour.

No. 13.—A smoky-brown carriage Horse dappled with black on the hind-quarters.

No. 14.—I have seen a roan Horse dappled all over with *black* spots, much like the grey, dun, or brown dappled Horse, but usually both the roan

Z

and strawberry-roan Horses show only *vestiges* of spots, either black in the roan, or brown in the strawberry-roan.

No. 15.—The nearest approach I have yet seen to Leopard spots in the Horse was in a dark-brown Horse with lighter spots; if some parts were cut out, they might have been passed off as parts of a black Leopard.

No. 16.—A dun-coloured Horse, seven years old, in a hansom, had three or four Zebra stripes above his wrist, and similar, but fainter ones, above his heel.

No. 17.—A dappled grey Horse had black lines across the upper part of fore-legs, corresponding to the transverse veins; a little more decision in the colour might easily turn them into Zebra stripes.

No. 18.—A white Pony had his mouth, circles round his eyes, and circlets just above the four hoofs, all of a yellow or golden bay colour.

No. 19.—A dun-coloured omnibus Horse had a black line down his spine, and two narrow stripes on his withers.

No. 20.—A bay Horse with white blotches and spots somewhat similar to those of Fig. 37.

No. 21.—A bay or black Horse with a white mane and tail is not common, but sometimes it is met with; dun and cream-coloured Horses with a white mane and tail are more common.

No. 22.—A brown omnibus Horse with black spots.

No. 23.—A jet-black Horse, of the kind used in funerals, had no vestige of spots, but its hind-feet were *white*. It is rare to see jet-black funeral Horses with traces of spots on their hindquarters, but I have seen some which had them.

No. 24.—A strawberry-roan omnibus Horse, *newly clipped*, showed closely-set rosettes like those of the Jaguar (see Fig. 56 (a)), consisting of a largish dapple surrounded by a circle of small spots. This was the most satisfactory example of true rosettes in the Horse. Others of a similar character are sometimes seen here and there on the Horse. Usually the rosettes are *fused* into dapples.

No. 25.—In some dark-grey Horses the white spots are almost effaced, and are as faint as the spots on the legs of adult Lions.

No. 26.—A bay Horse with faint black spots.

No. 27.—A white carriage Horse, spotted black on its shoulder, neck, and fore-legs. The spots were not unlike those of the Cheetah. On the head it had brown blotches.

No. 28.—Now and again a light bay Horse is seen with true rosettes on its flank, almost exactly like those of a Leopard, but close together, and with all the component spotlets distinct.

No. 29.—A black carriage Horse with a black star on a white blaze.

No. 30.—A costermonger's Donkey, of a light-grey colour, with faint darker *spots* on its flank, such as many Horses have. This and the following are the only Donkeys I have seen which had any indication of having descended from a spotted ancestry.

No. 31.—A grey Donkey with large dark dapples.

No. 32.—A costermonger's brown Donkey had a white abdomen, a white mouth, and white circles round its eyes.

APPENDIX D

THE innumerable mammals which are fully or vestigially spotted or striped, can be seen in Museums and Zoological Gardens, but it may not be generally known that many marsupials give indication of having descended from a spotted or striped ancestry, as the following list, taken from Gould on the *Mammals of Australia*, will amply show :—

No. 1.—*Antechinus maculatus* has white spots on the abdomen only.
No. 2.—*Dasyurus maculatus* is brown, spotted with white on the body, and also on the tail.
No. 3.—*Dasyurus viverrinus* (variable Dasyure), is either black or olive colour, spotted with white on the body only. Other Dasyures are also spotted.
No. 4.—*Halmaturus Derbyanus* (Derby's Wallaby) is spotted.
No. 5.—A variety of *Petrogale xanthopus* (yellow-footed rock Wallaby) is faintly spotted and faintly ring-tailed also.
No. 6.—*Lagorchertes Leichhardti* is spotted.
There are other Marsupials which are variously marked :—
No. 7.—Wombats, such as *Phascolomys*, have faint spots in various parts.
Then the following are very distinctly marked :—
No. 8.—*Parameles Gunnii*, banded transversely.
No. 9.—*Tarsipes rostratus* has three longitudinal bands.
No. 10.—*Myrmecobius fasciatus* is banded like a Zebra. *Thylacinus cynocephalus* (Tasmanian Wolf),—see Appendix A, No. 7,—has broad black bands.
No. 11.—*Macropus fuliginosus* (Sooty Kangaroo) has vestiges of stripe on various parts ;
No. 12.—*Lagorchertes fasciatus* (banded Hare-Kangaroo) is banded like a Zebra on hindquarters, with a spinal band, and it has faint spots on other parts.

APPENDIX E

I CONSIDER that a ringed tail, even where no other vestige of ancestral marking exists, is sufficient indication that the possessor of it descended from either a rosetted, or a spotted, or a striped ancestor. This is one of those features of which it cannot be said, when it exists *alone* as marking, that it was brought about by natural selection to harmonise with the surroundings of the animal, as a plain body with only the *tail* mottled would be too great an absurdity in defence of harmonisation.

The following list will give some idea of the number of *different* animals which have ringed tails :—

No. 1.—All Leopards, Tigers, the Cheetah, and all spotted and striped Cats, have ring-tails.

No. 2.—A number of Civets and Genets, whose bodies are either spotted or striped, have ring-tails.

No. 3.—The Delundung (*Prionodon gracilis*) and *Hemigalea Hardwickii* are ring-tailed.

No. 4.—A Hill Fox (*Vulpes montanus*) is brownish-yellow with faint rings on its tail. I saw another ring-tailed Fox in the Durham Museum.

No. 5.—The red-armed Squirrel of Fernando Po (*Sciurus rubi-brachiatus*) has a grey-brown body and a ringed tail.

No. 6.—The Grissled Squirrel (*Sciurus punctatus*) of West Africa has a ringed tail.

No. 7.—The Burmese Squirrel (*Sciurus pygerythrus*) is ring-tailed.

No. 8.—The Coati (*Nasua rufa*) and *Bassaris astuta* have ring-tails and plain bodies.

No. 9.—A Lemur, a Marmoset, and a spotted Ichneumon, of Jamaica, have ring-tails.

No. 10.—The African Linsang (*Poiana poensis*) is spotted and also ring-tailed.

No. 11.—The Pand (*Ælurus fulgens*) is ring-tailed.

No. 12.—The Racoon (*Procyon lotor*), and the Egyptian Cat, though quite plain, have ring-tails.

No. 13.—Burchell's Zebra and Chapman's Zebra have their tails ringed half-way.

Then if we turn again to Gould's *Australian Mammals* we find that:—

No. 14.—*Belidens scurus* has vestiges of rings on its tail. *Petrogale xanthopus* (yellow-footed rock Wallaby) is banded on the neck and has rings on its tail.

No. 15.—*Canis dingo* has three faint rings on its tail.

N. B.—These lists should not be considered as giving in any way all the animals that exist under those headings, nor are they meant for a scientific classification of the animals contained in them.

They are only intended to show the general reader the large number of mammals that are either spotted or striped, or are simply ring-tailed, which feature I consider a *vestige* of either spotted or striped ancestry. In these lists I have left out all domestic mammals, which are known to every one.

The probability is that even those mammals which are wholly plain at all ages, judging from domestic animals, descended from marked ancestors also, but from some cause they entirely lost their marks at an early period, and now the nervous system has lost the habit of reproducing them.

INDEX

Antelopes, contrasted coloration of, 141; markings of, 47-49.
Armadillo, why it has rosettes on abdomen, 191; hind-feet ungulate, 191; spots on abdomen not evolved by natural selection, 192; no excuse for rosettes on abdomen, 193; abdominal rosettes evidence of ancestral plate-armour, 194.
Armour-plates, of various animals, 111-115; as blocks of Jaguar rosettes, 121; round eyes of Ganoids, 180; early loss of, cause of contrasted coloration, 181; dwindling of, in certain animals, 202; gradual disappearance of, 205; disappearance, due to famine of lime-salts, 206.
Asses, striped and spotted, 89, 91.

Badgers, contrasted coloration of, 138.
Begonia, spotting on leaves of, 158.
Black Buck, contrasted coloration of, 138-141.
Blaze, of Horses and forehead star, 167-169; on face of other mammals, 171; possibly a vestige of ancestral forehead and nose-horns, 172.
Blotching: of Dog, 51; of Horse, 82; of Giraffe, 95.
Brain, development proportional to muscular activity, 210.
Bull, spotting of, 95, 96.

Callosities, loss of, on legs of ruminants, cause of distinct tufts of hair, 181-233; on metatarsal region of Tragulus, meaning of, 187, 232; on legs of Equidæ, 232-240.
Cheetah, analysis of markings, 20-24.
Chlamydophorus truncatus, transition mammal, 139; armour of, in transition stage, 204.
Coati, meaning of ringed tail of, 197-203.

Coloration of Mammals, Mr. Poulton's views, 105; Mr. Tylor's views, 107; Mr. Darwin's views regarding white marks, 108; not result of natural selection, 124; contrasted, 137-139; contrasted coloration, meaning of, 139-142, 181; of *Mellivora Indica*, 144; of *Putorius sarmaticus*, 144; of Spotted Deer, 145; of Skunks, 146, 147; of Jackal and other mammals, 147; of rosettes, 149; when contrasted, a vestige of a carapace, 151; contrasted coloration round mouth, eyes, and ears, meaning of, 175; has no hard and fast fixity, 178; of cattle in Cadzow and Somerford Parks, 179; striking, in Bear of Eastern Tibet, 183; of American Bear, 184.
Crocodile, strong armour on abdomen, 193-201.
Croft, W. B., experiments with impressions on glass, 160.

Dappling, of Horse, 60-69.
Deer, spotting of, 25, 46, 145; eye gland of, 237, 238.
Dogs, spotted, blotched, and striped, 50-53; Wolf ancestor of, 186.

Eimer, Professor, views on spotting and striping, 126.
Evolution in Mammals, four distinct stages of, 183.

Felidæ, in Elliot's monograph, 11.
Fishes, with armour round eyes, 180.
Flower, Sir W., on origin of Horse callosities, 232.
Foraminifera, hoarding of lime-salts by, 206.

360 STUDIES IN THE EVOLUTION OF ANIMALS

Giraffe, markings of, 95, 96; New, 98; blotching on abdomen not needed, 194.
Glass, retains impressions of coins and print, 160.
Glyptodonts, rosettes in armour of, 109-112.

Hands and feet, white or black, genesis of, 174.
Homology, of Rhinoceros shields with Armadillo shields, 215; of plates of *R. Sondaicus* with plates of Armadillo, 218-220; of hide-plates of Indian Rhinoceros with bone-plates of Crocodile and Sturgeon, 221; of pigment reticulation of Horse with commissures of plates of *R. Sondaicus*, 222.
Horses, dappling of, 59; want of data regarding commencement of dappling, 67; need of photographic help in establishing facts of dappling, 69; coincidence of venation and reticulation, 71; fern-like markings of, 72; detail study of rosettes of, 74-77; vestiges of Zebra marks in, 78-81; blotches in, 82; striped, 83-88; rosetted, 93; white or black hands and feet, explanation of, 174; tan-coloured circle round eye of, 177; limbs of ancestors of, 231; abnormal hands of, 251-253; are they odd-toed or even-toed, 252; characters in common with ruminants, 267.
Hybrids, between genera of plants, 173; between Onager and Abyssinian Ass, 170.

Interbreeding, of species and genera, 293, 294.

Jaguar, a South American Leopard, 6; black varieties of, 6; a tree-loving animal, 8; detail of rosettes of, 10; modified rosettes of, 18; localised nerve-centres of markings, 19; variants of rosettes of, 24; marking of, not result of natural selection, 124; coloration elements of, 162.

Leopards, identity with Panther, 7; black cubs of, in same litter with ordinary kinds, 7; Asiatic, differ from African, 8; tree-loving animals, 8; detail of rosettes of, 10-102; difference in markings of, 26; disposition of markings in, 27; spotting on legs of, 40; Clouded, 35.
Lime-salts, famine of, cause of disappearance of skin-armour, 207; immense quantities required for endo-skeletons of animals, 208; hoarded in lime-rocks, 206.
Lion, ancestral spotting of, 28.
Llama, callosities on hind-legs of, 236.
Lydekker, *Richard*, views on spotting and striping of mammals, 126.

Mammals, striped or spotted, 3; tail-rings of, 3-12; earliest record of striped, 4; markings of, supposed to be result of natural selection, 5; unsymmetry of markings in, 5; invisibility of, owing to markings, 6; markings of, dependent on nerve-centres, 26, 27; review of markings of, 54, 55; skeleton of, developed independently of coloration of skin, 126; contrasted coloration of certain, 137-139; white or black hands and feet in, explanation of, 174; contrasted coloration round mouth, eyes, and ears of, 175; four distinct stages in evolution of, 183; endo-skeleton of, required immense quantities of lime-salts, 208.
Marbled Cat, descent from rosetted ancestors, 30, 35.
Markings, of Jaguar, 9, 13, 18, 24; of Leopard, 9-15, 18, 27; of Cheetah, 20-23; of Spotted Deer, 25, 46, 145; of Lion, 29; of Puma, 29; of Serval, 30; of Marbled Cat, 31; of Ocelots, 32, 33; of Tigers, 36-41, 44; of Lynx, 40; of smaller Cats, 42, 43; of Zebras, 83, 84; on operculum of Pearly Nautilus, 120; of Jaguar, impressions of ancestral carapace, 120; meaning of, on fore-legs of Leopards, 124; of Jaguar not result of natural selection, 124; of fishes, 155; round mouth, eyes, and ears of certain mammals, 175; on hind-legs of Serval, 30, 186; on abdomen of Giraffe and Horse not needed, 194; of Leopards evidence of ancestral scutes, 202; order in which they altered in Mammals, 211.
Monstrosities, in various animals, 276-279, 287; rejection of, by Dr. Wallace as modifiers of species, 280; sub-division of, 282; probable cause of, 282-286; Geoffroy St. Hilaire's propositions on,

INDEX

289; possible explanation of, 294; digital, 295; in Guinea-pig, 302; in *Myosotis*, 304; wide range of monstrous species, 317; Hornbills as monstrous birds, 318; monstrous human hands and feet, 321; a help towards comprehending the doctrine of evolution, 323.

Natural selection, maintained but not originated skin-rosetting, 196, 197.
Nervous system, habits and memories of, alter very slowly, 227.

Ocelot, analysis of markings, 32-34.
Odontoglossum crispum, markings of, 24.
Orchids, persistent abnormality in sexual organs of, 255.
Order in which markings of mammals have altered, 211.
Ostracion, unarmoured lips of, 180.

Pangolins, unarmoured parts of, 141; scales of, agglutinated hairs, 159; characteristics of, 181.
Peaches and other fruit, colouring of, 139.
Pichiciago, rump shield of, 140; transition stage of armour of, 204.
Pig, abnormal hand of, 247.
Plates, armour, of various animals, 111-115; fusion of, in Glyptodon, 117; in Leathery Turtle, 118; restored in ancestral Jaguar, 122; on abdomen of Great Armadillo, 123; transverse on fore-legs of Hairy Armadillo, 124; of Dolphins, 133; originally both dorsal and ventral, 151; in fishes, 155-157; homology of, with pigment rosettes, 161; early loss of, round mouth, eyes, and ears, 175; influenced by nervous system, 121; disappeared gradually, 204-205; disappearance of, due to famine of lime-salts, 206, 207.
Polydactylism, cases of, 294-298; in Ichthyosaurs, 299.
Propithecus, contrasted coloration of, 138.
Puma, ancestral spotting of, 28.

Quagga, coloration of, 87-88.

Rabbit, meaning of white under-surface of tail, 143.
Racoon, ringed tail of, vestige of ancestral body-spotting, 28, 203.

Reason, why huge extinct animals have had pigmy descendants, 209; of rachitic egg-skin, 210.
Relationship, of markings of Horse, Giraffe, and Zebu, to plating of Armadillo and Rhinoceros, 215, 224.
Rosettes, of Jaguar, 9, 14, 18, 24, 75; of Leopard, 4, 9, 15-18, 27; of Ocelots, 32, 33; of Horse, 75, 76, 93; of Zebu, 95; of Jaguar, meaning of, 109; variations of, in Jaguar and Leopard, 102; bone-rosettes of Glyptodonts, 109, 112; of *Polacanthus*, 112; of Leathery Turtle, 118; of Crocodile and Sturgeon, 118; of extinct fishes, 119; of Jaguar not result of natural selection, 124; rosettes and spots have deeper meaning than general colour, 159; why on abdomen of Armadillo, 191; on abdomen of Leopard, why more altered than on back, 201.
Ruminants, endo-skeletons of, absorbed immense quantities of lime-salts, 208; contrasts of hair-tufts on legs of, evidence of ancestral callosities, 233, 234; characters of, in common with Horses, 267.

Scutes, dermal, disappeared gradually, 204, 205.
Selous, F. C., on markings of Antelopes, 47; hiding in tall grass, 238; on Antelopes living in waterless regions, 48.
Serval, origin of markings of, 30; meaning of black marks from heel to toes, 30; melanoid varieties of, 11.
Serranus gigas, spotting of, 156; other species of, with vestiges of plate-armour, 157.
Sheep, fleece of, what it tells us, 144; Cyprian, white mouth of, 177.
Skeleton, of mammals developed independently of skin-coloration, 126.
Skunks, coloration of, 146, 147.
Spinal line, of certain mammals, possible meaning of, 140.
Spotted Deer, markings of, 25; what they tell us, 144.
Spotting, of Cheetah, 20, 23; of Deer, 25, 46, 145; of Lion, 28; of Puma, 28; of Serval, 30; of Marbled Cat, 30; of Dog, 50; of Horse, 77; of Bull, 95; over eyes of black-and-tan Dog, meaning of, 176.

Striping, of Tigers, 36, 41, 44; of small Cats, 42, 43; of Dog, 52; of Horse, 77, 80; of Zebras, 83, 88; of Antelopes, 47.
Supernumerary digits, cases of, 294-298.
Survivals, of wholly armoured, of partly armoured, and of unarmoured stages of animals, 204, 205.
Symmetry, want of, on both sides of animals, 283.

Tails, ringed, 35, 43; of Coati, 183; evidence, when ringed, of ancestral rosetted or striped body, 197; of Racoon, 203.
Teratogeny, experiments in, 290.
Tiger, analysis of markings in, 35, 39, 40; extraordinary ocelli on skin of, 41, 44.
Types, new, inroad of, in geological times, 307-311; speculations as to possible causes of, 311.

Veins, reticular on Horse flank, equivalents of commissures in armour of *K*.
Sondaicus, 225.
Vertebrates, comparison of limbs of, 249, 257, 261, 264.
Vestiges, of ancestral features, 178.
Viverridæ, markings of, 53.

Werner, Franz, researches on spotting of animals, 73.
Wolf, ancestor of Dog, 186.

Zebra, variations in striping of, 83, 88; coloration of feet of Chapman's, 179; stripes on abdomen not needed, 194; Grevy's, wholly without callosities, 85, 234.
Zebu, rosetting of, 95, 97.
Zeuglodonts, armour-plating of, 133.

www.ingramcontent.com/pod-product-compliance
Lightning Source LLC
Chambersburg PA
CBHW032010220426
43664CB00006B/203